W9-ADN-630

Lecture Notes in Mathematics   1870

Editors:
J.-M. Morel, Cachan
F. Takens, Groningen
B. Teissier, Paris

Vladimir I. Gurariy · Wolfgang Lusky

Geometry of Müntz Spaces and Related Questions

 Springer

Authors

†Vladimir I. Gurariy
Department of Mathematics
Kent State University
Kent OH 44242
U.S.A.

Wolfgang Lusky
Institute of Mathematics
University of Paderborn
Warburger Str. 100
D-33098 Paderborn
Germany
e-mail: lusky@uni-paderborn.de

Library of Congress Control Number: 2005931756

Mathematics Subject Classification (2000): 46B20, 46B15, 46E15, 51A10

ISSN print edition: 0075-8434
ISSN electronic edition: 1617-9692
ISBN-10 3-540-28800-7 Springer Berlin Heidelberg New York
ISBN-13 978-3-540-28800-8 Springer Berlin Heidelberg New York

DOI 10.1007/11551621

Springer is a part of Springer Science+Business Media
springeronline.com
© Springer-Verlag Berlin Heidelberg 2005
Printed in The Netherlands

Typesetting: by the authors and TechBooks using a Springer LaTeX package

Cover design: design & production GmbH, Heidelberg

Printed on acid-free paper   SPIN: 11551621   41/TechBooks   5 4 3 2 1 0

After the completion of our book the first named author, Vladimir I. Gurariy, died. The world lost a great mathematician and I lost a close friend.

Wolfgang Lusky

Preface

Let $\Lambda = \{\lambda_k\}_{k=0}^{\infty}$ be an increasing sequence of non-negative numbers:

$$0 = \lambda_0 < \lambda_1 < \lambda_2 < \ldots$$

Moreover let $M(\Lambda) = \{t^{\lambda_k}\}_{k=0}^{\infty}$ be the sequence of the functions t^{λ_k} on $[0,1]$ and let $[M(\Lambda)]_E$ be the closed linear span of $M(\Lambda)$ in a given Banach space E containing $M(\Lambda)$. We call $M(\Lambda)$ a Müntz sequence and $[M(\Lambda)]_E$ a Müntz space.

In our book we shall be mainly concerned with $E = C := C[0,1]$, the Banach space of all realvalued continuous functions on $[0,1]$ endowed with the sup-norm, and $E = C_0 := C_0[0,1]$, the subspace of C consisting of all those functions $f \in C$ with $f(0) = 0$. Furthermore we deal with $E = L_p = L_p[0,1]$, $1 \leq p \leq \infty$, the space of all (classes of) realvalued measurable functions f on $[0,1]$ with

$$\|f\|_{L_p} = \left(\int_0^1 |f(t)|^p dt \right)^{1/p} < \infty \quad \text{if } 1 \leq p < \infty .$$

If $p = \infty$ then we take for $\|f\|_{L_\infty}$ the essential sup-norm instead.

We want to study geometric properties of the corresponding Müntz sequences and spaces. Let us begin with the famous Müntz theorem, [110]:

For $E = C$ or $E = L_p$, $1 \leq p < \infty$, we have

$$[M(\Lambda)]_E \neq E \quad \text{if and only if} \quad \sum_{k=1}^{\infty} \frac{1}{\lambda_k} < \infty .$$

(A proof of this fact in more generality will be given in 6.1.)

So, if $\sum_{k=1}^{\infty} 1/\lambda_k < \infty$ we obtain new Banach spaces $[M(\Lambda)]_E$. This sets the stage for the central problem we discuss in (Part II of) our book:

What kind of Banach space $[M(\Lambda)]_E$ do we obtain depending on the given Λ if $\sum_{k=1}^{\infty} 1/\lambda_k < \infty$?

This problem is far from being solved. Here we present the known theorems and prove new results in this direction. For example, if Λ is quasilacunary then $[M(\Lambda)]_{L_p}$ is isomorphic to l_p for $1 \leq p < \infty$ and $[M(\Lambda)]_C$ is isomorphic to c_0 (Sect. 9.1). But for non-quasilacunary Λ this is not always the case. There are at least two different isomorphism classes for $[M(\Lambda)]_C$ (Sect. 10.2). Moreover there is a continuum of different isometry classes for $[M(\Lambda)]_C$ (Sect. 10.4). In general, $[M(\Lambda)]_E$ can be regarded as a sequence space rather than a function space. $[M(\Lambda)]_{L_p}$ is always isomorphic to a subspace of l_p and $[M(\Lambda)]_C$ is isomorphic to a subspace of c_0 provided that the Müntz condition $\sum_k 1/\lambda_k < \infty$ and the gap condition $\inf_k(\lambda_{k+1} - \lambda_k) > 0$ are satisfied. In addition, $[M(\Lambda)]_{L_1}$ is always isomorphic to a dual Banach space (Sect. 9.1).

It is an open problem if every $[M(\Lambda)]_E$ has a basis. We discuss more general bounded approximation properties in Chap. 9. However, $[M(\Lambda)]_C$ can never have a monotone basis (Sect. 9.4). In this context it is interesting to note that $M(\Lambda)$ is always a minimal system provided that the Müntz condition holds. But $M(\Lambda)$ is never a basis or even uniformly minimal in $[M(\Lambda)]_E$ for $E = C$ or $E = L_p$ unless Λ is lacunary (Sect. 9.3). In contrast to Müntz sequences the trigonometric system $\{z^k\}_{k=-\infty}^{\infty}$ on $\{z \in \mathbf{C} : |z| = 1\}$ is uniformly minimal and even an Auerbach system. The traditional bridge between the trigonometric system and the classical Müntz system $\{t^n\}_{n=0}^{\infty}$, the substitution by Chebyshev polynomials [12], breaks down if we go over to subsequences of $\{t^n\}_{n=0}^{\infty}$. So there is no way to relate a general Müntz sequence $\{t^{\lambda_n}\}_{n=0}^{\infty}$ to the trigonometric system.

It is even unknown in general if the finite dimensional Müntz spaces $[M(\{\lambda_0, \lambda_1, \ldots, \lambda_n\})]_C$ have uniformly bounded basis constants. In Sects. 10.3 and 12.2 we discuss some special cases and related questions. In Chap. 12 we investigate phenomena which, we feel, deserve further investigation. Take a Müntz sequence $\{t^{\lambda_k}\}_{k=1}^{\infty}$, fix n and put $B_m = \text{span}\{t^{\lambda_{m+1}}, \ldots, t^{\lambda_{m+n}}\}$ in C. Then, for many different $\Lambda = \{\lambda_k\}_{k=1}^{\infty}$, the sequence of n-dimensional Banach spaces $\{B_m\}_{m=1}^{\infty}$ converges to the subspace $\text{span}\{t, t \log t, \ldots, t^{n-1} \log t\}$ of C with respect to the logarithm of the Banach-Mazur distance. This might be helpful for gaining further insight in the isomorphism character of $[M(\Lambda)]_C$. In Chap. 11 we treat more general classes of subspaces of $C[0,1]$ which have many common features with $[M(\Lambda)]_C$.

It is well-known that there is a close relationship between the theory of Müntz spaces and fields like approximation theory, harmonic analysis and functional analysis. The first major contribution to this theory after the seminal papers of Müntz [110] and Szász [136] was given by L. Schwartz [128] and Clarkson and Erdős [19] who established the fact that, for integer Λ, each $x(t) \in [M(\Lambda)]_C$ has an analytic continuation to the open complex unit disk. This means, for example, that $[M(\Lambda)]_C$ consists entirely of functions which are real-analytic on $]0, 1[$ provided that the Müntz condition and the gap condition hold! (See Sect. 6.2.)

In our book we want to change the accent from an analytical to a more geometrical approach and attempt to put well-known and new results into the perspective of a geometrical framework. At the same time we do not pretend completeness, we rather want to put the emphasis on unsolved problems, conjectures and ideas according to the taste of the authors. Although there is a natural overlap in this book with portions from excellent books such as [12] and [22] we present this material here from our geometric point of view. It seems to be the first time that Müntz spaces are treated under strict geometric orientation.

We assume that the reader has a basic knowledge of functional analysis.

The book is divided into two parts and twelve chapters. The first part contains the preliminary material from the geometry of normed spaces which is then applied to concrete Müntz spaces in Part II and which the authors believe to be promising for further investigation.

Both parts are essentially selfcontained and can be read independently of each other. In the summary Part I we skip some of the proofs and refer to the literature instead while, as a rule, in Part II we work out the proofs in full detail.

But Part I is more comprehensive than necessary for a simple outline of the preliminaries to Part II. There we give a systematic treatise of classical Banach space notions such as opening and inclination of subspaces (in Chap. 1). Moreover we introduce the projection function and projection type of a Banach space (1.6) and discuss their relation to Banach spaces with or without bases. Here the study of dispositions of subspaces in Banach spaces plays the main role.

In Chap. 2 we deal with general sequences in Banach spaces and properties such as minimality, completeness or stability. After the introduction of basic notions such as isomorphisms and the Banach-Mazur distance in Chap. 3 we study spaces which are (almost) universal with respect to a given class of Banach spaces and similar notions for bases in Chap. 4. Finally, Chap. 5 is devoted to a discussion of approximation properties centered around the commuting bounded approximation property (CBAP).

All our Banach spaces are assumed to be real unless indicated otherwise. (But almost all proofs in the following can be taken over literally to the complex case.) If E is a Banach space let E^* denote its topological dual space, i.e. the space of all linear bounded functionals on E.

Kent, Paderborn
May 2005

Vladimir I. Gurariy[1]
Wolfgang Lusky

---

[1] Supported by Deutsche Forschungsgemeinschaft

Contents

Part I

Subspaces and Sequences in Banach Spaces

In the first part of our book we will be concerned with dispositional properties of a Banach space E, i.e. the geometry of E and its subspaces. They are important tools for the analysis of E and the study of phenomena such as basic decompositions, approximations etc. A special role is played by the connection between "dispositional" properties (in terms of angles, etc.) and "distancional" properties (in terms of Banach-Mazur distance) which were discovered and developed in the sixties, [43, 82, 116].

In Chap. 1 we introduce the basic notions of the subspace disposition theory (see [43]) while in Chap. 2 we deal with applications to sequences in normed spaces. We will mention some often used technical theorems in the spirit of "planimetry" or "stereometry" in normed spaces.

The emphasis of Chap. 3 lies on isomorphisms and embeddings of Banach spaces. In Chap. 4 we study Banach spaces with almost universal disposition. Finally, Chap. 5 is devoted to bounded approximation properties properties of normed spaces involving various operators of finite rank (FDD, CBAP, etc), see [99–101].

So Part I is related to questions which, besides being of independent interest, also will lead us to the study of the geometry of Müntz and Müntz-type sequences in Part II.

1

Disposition of Subspaces

In this chapter we discuss how two or more subspaces in a Banach space affect each other by their position in a Banach space and we give applications in the geometry of Banach spaces.

We start with a discussion of well-known different definitions of opening and relate these notions to the inclination of subspaces. This leads, for example, to conditions for the closure of the sum of two subspaces. Finally we focus on operator theoretic aspects. We introduce projection constants and discuss the notions of load and projection function which turn out to be important tools for the analysis of a Banach space.

1.1 Different Definitions of the Opening of Subspaces

M. Krein, M. Krasnoselskii and D. Milman introduced in [74] the following definition of the opening of two subspaces U and V in a Banach space E:

$$\hat{\Theta}(U,V) = \max \left\{ \sup_{x \in U, ||x||=1} \rho(x,V), \sup_{y \in V, ||y||=1} \rho(y,U) \right\}.$$

(Here $\rho(\cdot,\cdot)$ denotes the distance with respect to the metric given by the norm.)

A significant part of the applications of this concept is based on the following theorem proved in [74] (see also [44]). Recall, the *density character* of a Banach space E is the smallest cardinality of a dense subset of E.

Theorem 1.1.1 *Assume that, for the subspaces U and V of E, one of the following conditions holds:*
(i) The density characters of U and V are different.
(ii) One of the spaces U and V is infinite dimensional and the other one is finite dimensional.
(iii) Both spaces are finite dimensional and their dimensions do not coincide. Then we have

$$\hat{\Theta}(U,V) \geq 1/2 .$$

If in addition at least one of the subspaces U and V is finite dimensional or E is a Hilbert space then $\hat{\Theta}(U,V) = 1$.

If E is a Hilbert space and dim $U = $ dim V then $\hat{\Theta}(U,V)$ can be quite small (see 1.3.1).

   I. Gohberg and A. Marcus [32] changed the definition of sphere opening by introducing the spherical opening $\tilde{\Theta}$ in the following way:

$$\tilde{\Theta}(U,V) = \max \left\{ \sup_{x \in S_U} \rho(x, S_V), \ \sup_{y \in S_V} \rho(y, S_U) \right\} ,$$

where S_U is the unit sphere in U, i.e. the set of all elements $x \in U$ with $||x|| = 1$. Analoguously, S_V is defined. They established the following

Theorem 1.1.2 *The set of all closed subspaces in a Banach space E is a complete metric space with respect to the spherical opening $\tilde{\Theta}(U,V)$ as metric.*

It is easy to see that

$$\hat{\Theta}(U,V) \leq 1 \quad \text{and} \quad \hat{\Theta}(U,V) \leq \tilde{\Theta}(U,V) \leq 2\hat{\Theta}(U,V) .$$

(Use the fact that $\rho(x,V) \leq \rho(x,S_V) \leq 2\rho(x,V)$ and $\rho(x,U) \leq \rho(x,S_U) \leq 2\rho(x,U)$ for $||x|| = 1$.) Theorem 1.1.1 and, accordingly, 1.1.2 become incorrect if one replaces $\hat{\Theta}$ by $\tilde{\Theta}$ and $\tilde{\Theta}$ by $\hat{\Theta}$, resp.

Examples. a) Take $\mathbf{R}^2$ with the Euklidean norm and put $U = \{(x,0) : x \in \mathbf{R}\}$ and $V = \mathbf{R}^2$. Then, according to 1.1.1, $\hat{\Theta}(U,V) = 1$ but $\tilde{\Theta}(U,V) = \sqrt{2}$.
b) Now let $E = \mathbf{R}^2$ be endowed with the norm $||(x,y)|| = \max(|x|, |y|)$. Put

$$U = \left\{ \left(x, \frac{x}{4}\right) : x \in \mathbf{R} \right\}, \quad V = \left\{ \left(x, \frac{x}{2}\right) : x \in \mathbf{R} \right\}$$

and $W = \{(x,0) : x \in \mathbf{R}\}$. An elementary computation shows $\hat{\Theta}(W,U) = 1/4$, $\hat{\Theta}(W,V) = 1/2$ and $\hat{\Theta}(U,V) = 1/6$. Hence

$$\hat{\Theta}(W,U) + \hat{\Theta}(U,V) < \frac{1}{2} = \hat{\Theta}(W,V)$$

which proves that $\hat{\Theta}$ is not a metric.

   Let us introduce a third definition of opening for which the statements of both theorems are correct.

Definition 1.1.3 *[43] The* ball opening *of the subspaces U and V in a Banach space E is defined by the following quantity*

$$\Theta(U,V) = \max \left\{ \sup_{x \in B_U} \rho(x, B_V), \ \sup_{y \in B_V} \rho(y, B_U) \right\} ,$$

where B_U is the unit ball in U, i.e. the set of all elements $x \in U$ with $||x|| \leq 1$ and likewise for B_V.

We derive from the definitions

Lemma 1.1.4 *We have*

$$(a) \quad \hat{\Theta}(U,V) \le \Theta(U,V) \le 1 \qquad and$$
$$(b) \quad \Theta(U,V) \le \tilde{\Theta}(U,V) \le 2\Theta(U,V).$$

Proof. (a) is a direct consequence of the definitions.
(b): If $||x|| = 1$ and $y \ne 0$, then we obtain

$$||x - \frac{y}{||y||}|| \le ||x - y|| + |\,1 - ||y||\,| \le 2||x - y||\,.$$

Hence $\rho(x, S_V) \le 2\rho(x, B_V)$ and, similarly, $\rho(y, S_U) \le 2\rho(y, B_U)$. This implies
$\tilde{\Theta}(U,V) \le 2\Theta(U,V)$. For arbitrary $x \in B_U$ with $x \ne 0$ we have

$$\rho(x, B_V) = ||x||\rho\left(\frac{x}{||x||}, \frac{1}{||x||}B_V\right) \le \rho\left(\frac{x}{||x||}, S_V\right)$$

and similarly $\rho(y, B_U) \le \rho(y/||y||, S_U)$. Hence

$$\Theta(U,V) \le \tilde{\Theta}(U,V) \le 2\Theta(U,V)\,. \qquad \square$$

If E is a Hilbert space then $\rho(x, U)$ is the norm of the orthogonal projection of
x with kernel U. This implies that $\Theta(U,V) = \hat{\Theta}(U,V)$ for all closed subspaces
of Hilbert spaces.

Theorem 1.1.5 *For the ball opening $\Theta(U,V)$ the statements of both Theorems 1.1.1 as well as 1.1.2 are valid.*

Proof. The statement of Theorem 1.1.1 for $\Theta(U,V)$ follows from 1.1.1 and
the fact that $\hat{\Theta} \le \Theta \le 1$. To prove the assertion of Theorem 1.1.2 for the
ball opening $\Theta(U,V)$, in view of the inequalities in 1.1.4 (b), it is sufficient to
observe that, again, Θ is a metric, i.e. the triangle inequality

$$\Theta(U_1, U_3) \le \Theta(U_1, U_2) + \Theta(U_2, U_3)$$

is satisfied. This follows by direct verification. $\qquad \square$

1.2 Inclination

Now we discuss a related notion.

Definition 1.2.1 *[43] Let U and V be subspaces of a Banach space E such
that $U \ne \{0\}$. The* inclination *of U to V is defined by*

$$(\widehat{U,V}) = \inf_{x \in U, ||x||=1} \rho(x, V)\,.$$

If V is spanned by the element $x \in E$ we will use the notation $(\widehat{U, x})$ instead of $(\widehat{U, V})$. Then we speak of the inclination of U to x. Analoguously we define the inclination of an element to a subspace and the inclination of an element to an element of E.

Let $U \cap V = \{0\}$ and let $P : U + V \to U$ be the projection with $P(u + v) = v$ for all $u \in U$ and $v \in V$. Then we easily obtain $(\widehat{U, V}) = ||P||^{-1}$. Indeed

$$||P|| = \sup_{x \in U, y \in V} \frac{||x||}{||x + y||} = \left(\inf_{x \in U, y \in V} \frac{||x + y||}{||x||} \right)^{-1} = \frac{1}{(\widehat{U, V})} .$$

(This even includes the case of unbounded P where $(\widehat{U, V}) = 0$ and distinguishes inclination from many other definitions of "angle between two subspaces".)

The definition of inclination has wide applications in the theory of bases (see, for example [53]) which is mainly due to the following criterion proved by Grinblum in equivalent terms (see [35, 36]).

For a given sequence $\bar{e} = \{e_i\}_{i=1}^{\infty}$ of elements in a Banach space E let us denote by $L_{i,j}$ the span of $e_i, e_{i+1}, \ldots, e_j$.

Definition 1.2.2 *$\bar{e}$ is called* complete *in E if closed span $\{e_i\}_{i=1}^{\infty} = E$.*

$\bar{e}$ is called basis *of E if each $x \in E$ has a unique representation as $x = \sum_{i=1}^{\infty} \alpha_i e_i$ where the series converges in norm.*

For more details about these notions see Sects. 2.3–2.5

Theorem 1.2.3 *[35] Let $\bar{e} = \{e_i\}_{i=1}^{\infty}$ be a complete system in a Banach space E such that $e_k \neq 0$ for all k. Then the following are equivalent*
(i) $\bar{e}$ is a basis of E
(ii) There is some $\beta > 0$ such that

$$(L_{1,i}, \widehat{L_{i+1,j}}) \geq \beta > 0 \quad whenever \ i < j$$

(iii) There is a constant $\beta > 0$ such that, for all choices of α_k,

$$\beta \left\| \sum_{k=1}^{i} \alpha_k e_k \right\| \leq \left\| \sum_{k=1}^{j} \alpha_k e_k \right\| \quad whenever \ i < j$$

Proof. The equivalence between (ii) and (iii) follows from the definition of inclination and the remark following 1.2.1

(i) $\Rightarrow$ (iii): Let $x \in E$, say $x = \sum_{k=1}^{\infty} \alpha_k e_k$. Put $|||x||| = \sup_n ||\sum_{k=1}^{n} \alpha_k e_k||$. Then, by assumption, $||x|| \leq |||x||| < \infty$. An elementary computation shows that E is complete under $|||\cdot|||$. So the open mapping theorem yields a constant $\beta > 0$, independent of x, with $\beta|||x||| \leq ||x|| \leq |||x|||$. Taking $x = \sum_{k=1}^{j} \alpha_k e_k$ we obtain (iii).

(iii) $\Rightarrow$ (i): Using (iii) we see that

$$\left\{ x \; : \; x = \sum_{k=1}^{\infty} \alpha_k e_k \text{ for some } \alpha_k \text{ with norm converging series} \right\}$$

is a closed subspace of E. Since $\bar{e}$ is complete we obtain that every $x \in E$ has a representation of the form $x = \sum_{k=1}^{\infty} \alpha_k e_k$. This representation is unique. Indeed if $0 = \sum_{k=1}^{\infty} \alpha_k e_k$ then (iii) implies that $\alpha_k = 0$ for all k. Hence $\bar{e}$ is a basis of E. □

The supremum of all β in the preceding theorem will be called the *index* of the basis $\{e_i\}_{i=1}^{\infty}$ and denoted by $\gamma(\{e_i\}_{i=1}^{\infty})$.

We also want to define the index $\gamma(\{e_k\}_{k=1}^{\infty})$ for a general sequence in E:

$$\gamma(\{e_k\}_{k=1}^{\infty}) = \inf\{(\widehat{L_{1,i}, L_{i+1,j}}) \; : \; i < j \}.$$

The notion of inclination is non-symmetric, i.e. we have $(\widehat{U,V}) \neq (\widehat{V,U})$ in general. The following proposition gives the value of the "degree of non-symmetry".

Proposition 1.2.4 *Let U and V be non-zero subspaces of E. If $(\widehat{U,V}) = \delta$ then $(\widehat{V,U}) \geq (1+\delta)^{-1}\delta$. If $E = C[0,1]$ then this inequality is sharp for any $\delta \in \,]0,1]$.*

Proof. Let $y \in V$, $||y|| = 1$, $x \in U$. We shall evaluate $||x + y||$. Consider two cases:

Case 1. $||x|| \leq (1+\delta)^{-1}$. Then

$$||x + y|| \geq ||y|| - ||x|| \geq 1 - \frac{1}{1+\delta} = \frac{\delta}{1+\delta} \, .$$

Case 2. $||x|| > (1+\delta)^{-1}$. Then

$$||x + y|| \geq \rho(x,V) \geq (\widehat{U,V})||x|| \geq \frac{\delta}{1+\delta} \, .$$

The elements $x \in U$ and $y \in V$ are chosen arbitrarily. Therefore, we have

$$(\widehat{V,U}) = \inf_{x \in U, y \in V, ||y||=1} ||x + y|| \geq \frac{\delta}{1+\delta} \, .$$

To show that this inequality is sharp consider the following two functions in $E = C[0,1]$: $x(t) = 1$, and $y(t) = (1 - \delta) + 2\delta t$. We check directly that $(\widehat{x,y}) = \delta$ and $(\widehat{y,x}) = (1+\delta)^{-1}\delta$. Thus the theorem is proved. □

Proposition 1.2.4 and Definition 1.2.1 imply the following

Corollary 1.2.5 *If $(\widehat{U,V}) \geq \delta$ then, for each $x \in U$ and $y \in V$, we have*

$$||x + y|| \geq \max\left(\delta||x||, \frac{\delta}{1+\delta}||y|| \right)$$

In Hilbert spaces we have symmetry.

Theorem 1.2.6 *Let E be a Banach space with dim $E > 2$. Then we have $(\widehat{U,V}) = (\widehat{V,U})$ for any two non-zero subspaces U and V if and only if E is a Hilbert space.*

Proof. [42] At first assume $(\widehat{U,V}) = (\widehat{V,U})$ for any two non-zero subspaces U and V of E. Let Q be a three-dimensional subspace of E and P a two-dimensional subspace of Q. Obviously there is an element $x \in Q$ such that $(\widehat{x,P}) = 1$ and so, by assumption, $(\widehat{P,x}) = 1$. This means that the cylindrical surface with directional line S_P (the unit sphere in P) along x supports the unit ball B_Q of Q. Since P is arbitrary this implies that B_Q is an ellipsoid (see [6]). Hence Q is a Euclidean space which implies that E is a Hilbert space ([20], p. 151).

Conversely, let E be a Hilbert space and assume that P, Q are subspaces of E. Obviously, for any given $\epsilon > 0$, there are one-dimensional subspaces $P_1 \subset P$ and $Q_1 \subset Q$ such that

$$(\widehat{P,Q}) \geq (\widehat{P_1,Q_1}) - \epsilon$$

We clearly have that $(\widehat{P_1,Q_1}) = (\widehat{Q_1,P_1}) \geq (\widehat{Q,P})$. This implies

$$(\widehat{P,Q}) \geq (\widehat{Q,P}) - \epsilon$$

Since $\epsilon > 0$ was arbitrary we obtain $(\widehat{P,Q}) \geq (\widehat{Q,P})$. Similarly, we infer $(\widehat{Q,P}) \geq (\widehat{P,Q})$ which completes the proof of 1.2.6. $\square$

We remark that there exists a two-dimensional non-Euclidean space with symmetric inclination where the unit sphere is a hexagon [61].

Theorem 1.2.7 *Let $\{e_i\}_{i=1}^{\infty}$ be a complete sequence in a Banach space E with $e_i \neq 0$ for all i. Assume that $(\widehat{L_{1,i},e_{i+1}}) = \beta_i$, $i = 1, 2, \ldots$. If $\beta = \prod_{i=1}^{\infty} \beta_i > 0$ then $\{e_i\}_{i=1}^{\infty}$ is a basis in E and $\gamma(\{e_i\}_{i=1}^{\infty}) \geq \beta$.*

Proof. Let $x \in L_{1,i}$ and $y \in L_{i+1,j}$ such that $y = \sum_{k=i+1}^{j} \alpha_k e_k$. Then

$$
\begin{aligned}
\|x + y\| &= \left\| x + \sum_{k=i+1}^{j-1} \alpha_k e_k + \alpha_j e_j \right\| \geq \left\| x + \sum_{k=i+1}^{j-1} \alpha_k e_k \right\| \cdot (\widehat{L_{1,j-1}, e_j}) \\
&\geq \left\| x + \sum_{k=i+1}^{j-2} \alpha_k e_k \right\| \cdot (\widehat{L_{1,j-2}, e_{j-1}}) \cdot (\widehat{L_{1,j-1}, e_j}) \geq \ldots \\
&\geq \|x\| \beta_i \beta_{i+1} \ldots \beta_{j-1} \geq \|x\| \beta
\end{aligned}
$$

Since $(\widehat{L_{1,i}, L_{i+1,j}}) \geq \beta$ whenever $i < j$ Theorem 1.2.3 implies that the sequence $\{e_i\}_{i=1}^{\infty}$ is a basis of E. $\square$

A sequence $\{e_k\}_{k=1}^{\infty}$ of elements in a normed space will be called *normalized* if $||e_k|| = 1$ for all k.

Definition 1.2.8 *The normalized sequence* $\{e_i\}_{i=1}^{\infty}$ *in the Banach space E is called δ-minimal for some $\delta > 0$ if* $\rho(e_i, L_{1,i-1} + L_{i+1,\infty}) \geq \delta$, $i = 1, 2, \ldots$. *Here* $L_{j,\infty} = span\,\{e_j, e_{j+1}, \ldots\}$.

We shall discuss minimal sequences in more details in Sect. 2.2. Here we prove

Theorem 1.2.9 *[43] Every normalized basis* $\{e_i\}_{i=1}^{\infty}$ *with index γ is δ-minimal where* $\delta = (1 + \gamma)^{-1}\gamma 2$. *In particular we have, for any* $x = \sum_{k=1}^{\infty} \alpha_k e_k$,

$$|\alpha_k| \leq \frac{1+\gamma}{\gamma 2}||x||, \quad k = 1, 2, \ldots .$$

Proof. Let $x \in L_{1,i-1}$, $y \in L_{i+1,\infty}$. Using Proposition 1.2.4 we have

$$||e_i + x + y|| \geq ||e_i|| \cdot (e_i, \widehat{L_{1,i-1}}) \cdot (L_{1,i}, \widehat{L_{i+1,\infty}})$$
$$\geq ||e_i||\frac{\gamma}{1+\gamma}\gamma = \frac{\gamma 2}{1+\gamma},$$

which means that $\{e_i\}_{i=1}^{\infty}$ is δ-minimal with $\delta = \gamma 2(1 + \gamma)^{-1}$. □

Besides inclination sometimes other versions of an angle between subspaces are used. Usually all of them are equivalent. Here we mention some of them as examples.

For non-zero elements $x, y \in E$ we define an *angle between x and y* by

$$\varphi(x,y) = \left\|\frac{x}{||x||} - \frac{y}{||y||}\right\|,$$

and an *angle between subspaces U and V* by

$$\varphi(U,V) = \inf_{x \in U, y \in V} \varphi(x,y) .$$

It is easy to obtain the relation

$$(\widehat{U,V}) \leq \varphi(U,V) \leq 2(\widehat{U,V}) .$$

For a Hilbert space H we define the *cos-angle* as

$$C(x,y) = \frac{|\langle x,y \rangle|}{||x|| \cdot ||y||} \quad \text{and} \quad C(U,V) = \sup_{x \in U, y \in V} C(x,y) .$$

For a sequence $\bar{e} = \{e_i\}_{i=1}^{\infty}$ it is natural to introduce the *angle index* as

$$\varphi(\bar{e}) = \inf_{n < m} \varphi(\text{span}\{e_i\}_{i=1}^{n}, \text{span}\{e_i\}_{i=n+1}^{m})$$

and *cos-index* as

$$C(\bar{e}) = \sup_{n<m} C(\mathrm{span}\{e_i\}_{i=1}^n, \mathrm{span}\{e_i\}_{i=n+1}^m)$$

In connection with the notion of angle we will introduce in Chap. 12 the concept of angle-convergence for a sequence of elements in a Banach space.

We conclude this section with an extension of the definition of inclination.

Definition 1.2.10 *Let A and B be two subsets of E such that $x/\|x\| \in A$ whenever $x \in A$ and $x \neq 0$. Then the* inclination *of A to B is defined to be*

$$(\widehat{A,B}) = \inf\{\rho(x,B) \; : \; x \in A, \; \|x\| = 1\}\,.$$

Note that this definition does not contradict 1.2.1 in the case that $B = \{x\}$ and A is a subspace. We distinguish between $(\widehat{A,\{x\}})$ and $(\widehat{A,x})$, i.e. the inclination of a subspace A and a singleton $\{x\}$ and the inclination of A and $\mathrm{span}\{x\}$.

1.3 Connection between the Opening and Inclination

The following inequality is obvious

$$(\widehat{U,V}) \leq \Theta(U,V)\,.$$

In a Hilbert space we sometimes have equality. For example, if U and V are orthogonal then $(\widehat{U,V}) = \Theta(U,V) = 1$. If U and V coincide then we always have $(\widehat{U,V}) = \Theta(U,V) = 0$. It turns out that in a Hilbert space, for any p with $0 \leq p \leq 1$ one can find subspaces U and V with dimensions larger than one which satisfy

$$(\widehat{U,V}) = \Theta(U,V) = p\,.$$

This is a particular case of the following

Theorem 1.3.1 *[43] In every infinite dimensional Hilbert space, for any integer $n > 1$ and numbers p, q with $0 \leq p \leq q \leq 1$ there exist n-dimensional subspaces U and V such that*

$$(\widehat{U,V}) = p \quad and \quad \Theta(p,q) = q\,.$$

Assuming $p = q$ we obtain

Corollary 1.3.2 *In every infinite dimensional Hilbert space, for every number p with $0 \leq p \leq 1$ and integer $n > 1$, there exist n-dimensional subspaces U and V such that*

$$\frac{\rho(x,V)}{\|x\|} = \frac{\rho(y,U)}{\|y\|} = p \quad for\ any\ x \in U,\ y \in V\ with\ x \neq 0 \neq y\,.$$

Proof. Take U and V as in 1.3.1 for $p = q$. We have, for $x \in U$ with $x \neq 0$,

$$p = (\widehat{U, V}) \leq \frac{\rho(x, V)}{||x||} \leq \Theta(U, V) = p$$

and hence equality. Similarly we obtain $\rho(y, U)/||y|| = p$ for $y \in V$.   □

Theorem 1.3.1 is also true for infinite dimensional subspaces.

The following indicates the continuous dependence of the inclination on some parameters, [57].

Lemma 1.3.3 *Let U, V and $\tilde{U}$, $\tilde{V}$ be non-zero subspaces in a Banach space E such that $(\widehat{U, V}) \geq \alpha > 0$ and $\Theta(U, \tilde{U}) < \delta$, $\Theta(V, \tilde{V}) < \delta$. Then*

$$(\widehat{U, \tilde{V}}) > \alpha - 2\delta \quad and \quad (\widehat{\tilde{U}, V}) > \alpha - 2\delta \ .$$

Proof. Take $x \in U$ with $||x|| = 1$ and take $z \in \tilde{V}$. Then we obtain

$$||x - z|| \geq \rho(x, V) - \rho(z, V) \geq (\widehat{U, V}) - \hat{\Theta}(V, \tilde{V})||z||$$
$$\geq (\widehat{U, V}) - \Theta(V, \tilde{V})||z|| > \alpha - \delta||z|| \ .$$

This implies $(\widehat{U, \tilde{V}}) > \alpha - 2\delta$. $(\widehat{\tilde{U}, V}) > \alpha - 2\delta$ can be proved similarly.   □

As a consequence we have

Theorem 1.3.4 *The inclination $(\widehat{U, V})$ is a uniformly continuous function of the non-zero subspaces U and V with respect to the metric defined by the ball opening Θ.*

Proof. Let $\epsilon > 0$ and put $\delta = \epsilon/4$. Consider subspaces U, $\tilde{U}$, V and $\tilde{V}$ of E with $\Theta(U, \tilde{U}) < \delta$ and $\Theta(V, \tilde{V}) < \delta$. Put $\alpha = (\widehat{U, V})$ and use 1.3.3 to obtain $(\widehat{\tilde{U}, \tilde{V}}) > \alpha - 4\delta$. Hence $(\widehat{\tilde{U}, \tilde{V}}) - (\widehat{U, V}) > -\epsilon$. Putting next $\alpha = (\widehat{\tilde{U}, \tilde{V}})$ and applying 1.c.3. again we see that $(\widehat{U, V}) - (\widehat{\tilde{U}, \tilde{V}}) > -\epsilon$. Thus $|(\widehat{U, V}) - (\widehat{\tilde{U}, \tilde{V}})| < \epsilon$.   □

Corollary 1.3.5 *The distance $\rho(x, V)$ from an element $x \in E$ to a subspace $V \subset E$ is a continuous function in x and V.*

Proof. Fix x and V and take $x_n \in E$, $V_n \subset E$ such that $\lim_{n \to \infty} ||x - x_n|| = 0$ and $\lim_{n \to \infty} \Theta(V, V_n) = 0$.

Case $x = 0$: Here we have $\rho(x, V) = 0$ and

$$\rho(x_n, V_n) \leq ||x_n - x|| + \rho(x, V_n) \leq ||x_n - x|| + \Theta(V, V_n) \ .$$

This shows that $\lim_{n \to \infty} \rho(x_n, V_n) = 0 = \rho(x, V)$.

Case $x \neq 0$: Then $x_n \neq 0$ for large n. Put $U = \text{span}\{x\}$ and $U_n = \text{span}\{x_n\}$. Since $x \neq 0$ we obtain $\lim_{n \to \infty} \Theta(U, U_n) = 0$. We have $\rho(x, V) = ||x||(\widehat{U, V})$ and $\rho(x_n, V_n) = ||x_n||(\widehat{U_n, V_n})$. Thus, by 1.3.4, $\lim_{n \to \infty} \rho(x_n, V_n) = \rho(x, V)$.   □

1.4 Conditions for the Closure of Sums of Subspaces

Now we apply the preceding notions. We consider

Definition 1.4.1 *Let U and V be closed subspaces of the Banach space E. If $U \cap V = \{0\}$ and $U + V$ is closed then $U + V$ is called* direct sum *and denoted by $U \oplus V$.*

If $U \cap V = \{0\}$ and $U + V$ is not closed then $U + V$ is called quasidirect sum *and denoted by $U \mathbin{\hat{+}} V$.*

It is obvious that $U+V$ is closed if U and V are finite dimensional. For infinite dimensional U and V the sum $U + V$ is not always closed. It is possible to reformulate the well-known geometric closure criteria of direct sums (see, for example, [36]) in terms of inclinations in the following way.

Theorem 1.4.2 *Let U, V be closed subspaces of a Banach space such that $U \cap V = \{0\}$. Then $U + V$ is closed if and only if $(\widehat{U,V}) > 0$.*

Proof. Let $P : U + V \to U$ be the projection with $P(x + y) = x$ for every $x \in U$ and $y \in V$. From the definition of inclination and the remark after 1.b.1. we obtain that P is bounded if and only if $(\widehat{U,V}) > 0$. Then we have $\|P\| = (\widehat{U,V})^{-1}$.

So, if $U + V$ is closed then it is complete and the closed graph theorem tells us that P is bounded. Hence $(\widehat{U,V}) > 0$.

Conversely, if $(\widehat{U,V}) > 0$ then P is bounded and $(id - P)$ is bounded. Since U and V are complete the space $U + V$ must be complete and hence closed. $\qquad\square$

Before we begin to formulate the main results of this section we mention the following useful

Proposition 1.4.3 *Let U, U_1, U_2, V, V_1, V_2 be subspaces of a Banach space satisfying $U = U_1 + U_2$, $V = V_1 + V_2$ and*

$$(U_1, \widehat{U_2 + V}) = \alpha, \qquad (V_1, \widehat{V_2 + U}) = \beta, \qquad (\widehat{U_2, V_2}) = \gamma$$

for some $\alpha > 0$, $\beta > 0$, $\gamma > 0$. Then we have

$$(\widehat{U,V}) \geq \frac{\alpha\beta\gamma}{\alpha + \beta + \alpha\beta + \alpha\gamma + \beta\gamma}.$$

Proof. Let $u \in U$, $v \in V$, $u_i \in U_i$ and $v_i \in V_i$, $i = 1, 2$, be such that $u = u_1 + u_2$, $v = v_1 + v_2$ and $\|u\| = 1$. Moreover, let a be the root of the equation

$$a\frac{\alpha\beta}{\alpha + \beta} = (1 - a)\gamma - a, \quad \text{that is} \quad a = \frac{\gamma(\alpha + \beta)}{\alpha + \beta + \alpha\beta + \alpha\gamma + \beta\gamma}.$$

Observe that $0 \leq a \leq 1$. To estimate $||u + v||$ we consider two cases:
Case 1. $||u_1|| + ||v_1|| \geq a$. Here we have two subcases:
   Either $||u_1|| \geq a\beta/(\alpha + \beta)$. Then we obtain

$$||u + v|| = ||u_1 + (u_2 + v_1 + v_2)|| \geq ||u_1||\alpha \geq \frac{a\alpha\beta}{\alpha + \beta}$$

Or $||u_1|| < a\beta/(\alpha + \beta)$. This implies $||v_1|| \geq a - ||u_1|| > a\alpha/(\alpha + \beta)$ and

$$||u + v|| = ||v_1 + (u_1 + u_2 + v_2)|| \geq ||v_1||\beta \geq \frac{a\alpha\beta}{\alpha + \beta}$$

Since $0 \leq \alpha \leq 1$, case 1 yields

$$||u + v|| \geq \frac{a\alpha\beta}{\alpha + \beta} = \frac{\alpha\beta\gamma}{\alpha + \beta + \alpha\beta + \alpha\gamma + \beta\gamma}$$

Case 2. $||u_1|| + ||v_1|| < a$. Here we obtain $||u_2|| \geq 1 - ||u_1|| \geq 1 - a$ and

$$||u + v|| \geq ||u_2 + v_2|| - (||u_1|| + ||v_1||) \geq ||u_2||\gamma - a \geq (1 - a)\gamma - a$$

By the choice of a we have $||u + v|| \geq \alpha\beta\gamma/(\alpha + \beta + \alpha\beta + \alpha\gamma + \beta\gamma)$ in this case, too. In view of the definition of $\widehat{(U, V)}$ this yields Proposition 1.4.3   □

Proposition 1.4.4 *("Almost orthogonality") Let U be a finite dimensional subspace of an infinite dimensional Banach space E. For arbitrarily given positive $\beta < 1$ there exists a finite codimensional subspace V_β of E with $\widehat{(U, V_\beta)} > \beta$.*

Proof. Fix γ with $\beta < \gamma < 1$. Find $e_1^*, \ldots, e_n^* \in E^*$ of norm one such that

$$\sup_k |e_k^*(u)| \geq \gamma||u|| \quad \text{for all} \quad u \in U$$

Such elements exist since U is finite dimensional. Let $V \subset E$ be the intersection of all kernels of the e_k^*. Then V is finite codimensional in E. If $u \in U$ and $||u|| = 1$ then, for any $v \in V$, we have $||u - v|| \geq \sup_k |e_k^*(u)| \geq \gamma$. Hence $\widehat{(U, V_\beta)} \geq \gamma > \beta$.   □

Now we introduce essentially non-isomorphic Banach spaces.

Definition 1.4.5 *(a) We call two infinite dimensional Banach spaces U and V partially almost isometric (partially isomorphic, resp.) if for any $\eta > 1$ (if for some $\eta < \infty$, resp.) there exist infinite dimensional subspaces $U_\eta \subset U$, $V_\eta \subset V$ and isomorphisms $T : U_\eta \to V_\eta$ such that $\max\{||T||, ||T^{-1}||\} < \eta$.*
   If U and V are not partially almost isometric (partially isomorphic, resp.) then they are called essentially non-isometric (essentially non-isomorphic, resp.)
   (b) A bounded linear operator $T : U \to V$ is called strictly singular if, for no closed infinite dimensional subspace $E \subset U$, $T|_E$ is an isomorphism.

In the following we frequently consider the classical sequence spaces

$$c_0 = \{\{\alpha_n\}_{n=1}^{\infty} \; : \; \alpha_n \in \mathbf{R} \text{ for all } n, \; \lim_{n \to \infty} \alpha_n = 0\}$$

$$c = \{\{\alpha_n\}_{n=1}^{\infty} \; : \; \alpha_n \in \mathbf{R} \text{ for all } n, \; \lim_{n \to \infty} \alpha_n \text{ exists}\}$$

$$l_{\infty} = \{\{\alpha_n\}_{n=1}^{\infty} \; : \; \alpha_n \in \mathbf{R} \text{ for all } n, \; \sup_n |\alpha_n| < \infty\} \, ,$$

endowed with the norm $\|\{\alpha_n\}_{n=1}^{\infty}\| = \sup_n |\alpha_n|$, and, if $1 \le p < \infty$,

$$l_p = \left\{ \{\alpha_n\}_{n=1}^{\infty} \; : \; \alpha_n \in \mathbf{R} \text{ for all } n, \; \sum_{n=1}^{\infty} |\alpha_n|^p < \infty \right\}$$

endowed with

$$\|\{\alpha_n\}_{n=1}^{\infty}\| = \left(\sum_{n=1}^{\infty} |\alpha_n|^p \right)^{1/p} .$$

Take $e_k = \{\alpha_{k,n}\}_{n=1}^{\infty}$ with $\alpha_{k,n} = 0$ if $n \ne k$ and $\alpha_{n,n} = 1$. We sometimes refer to the corresponding sequence $\{e_k\}_{k=1}^{\infty}$ as the *unit vector basis* of c_0 or l_p (if $p < \infty$).

To illustrate Definition 1.4.5 let, for example, E be one of the spaces c_0 or l_p, $1 \le p < \infty$, and let U and V be infinite dimensional subspaces of E. Then U and V are partially almost isometric. l_p and l_r, for $1 \le p < \infty$, $1 \le r < \infty$, $p \ne r$, as well as l_p, for $1 \le p < \infty$, and c_0 are examples of essentially non-isomorphic (therefore essentially non-isometric) Banach spaces. All this follows from the fact that, for every infinite dimensional subspace U of E and for every $\eta > 1$ there is a subspace $E_\eta \subset U$ and an (onto-)isomorphism $T : E_\eta \to E$ with $\max(\|T\|, \|T^{-1}\|) < \eta$. ([84], I Propositions 2.a.1. and 2.a.2.)

We mention two more results in this direction without proofs.

Proposition 1.4.6 *[84] Every linear bounded $T : l_{p_1} \to l_{p_2}$, where $p_1 \ne p_2$, $1 \le p_1, p_2 < \infty$, is strictly singular.*

The following theorem is based on the fact that l_{p_1} and l_{p_2} are essentially non-isomorphic if $p_1 \ne p_2$, $1 \le p_1, p_2 < \infty$.

Theorem 1.4.7 *(See [25]) Every complemented subspace of $l_{p_1} \oplus l_{p_2}$ for $p_1 \ne p_2$, $1 \le p_1, p_2 < \infty$, is isomorphic either to l_{p_1} or l_{p_2} or $l_{p_1} \oplus l_{p_2}$.*

Using Proposition 1.4.3 one can prove the following theorem ([43], independently obtained by H.P. Rosenthal, [124], in terms of "totally incomparable" Banach spaces).

Theorem 1.4.8 *Let U and V be closed subspaces of a Banach space E such that $U \cap V = \{0\}$. Moreover assume that U and V are essentially non-isometric. Then $U + V$ is closed.*

Proof. (a) At first we claim the following.

Let $U_1 \subset U$ and $V_1 \subset V$ be finite codimensional subspaces. Then $(\widehat{U,V}) = 0$ if and only if $(\widehat{U_1,V_1}) = 0$. (Here the fact that U and V are essentially non-isometric is not needed. We only use $U \cap V = \{0\}$.)

To prove the claim we observe that, if $(\widehat{U_1,V_1}) = 0$, then $(\widehat{U,V}) = 0$ follows directly from the definition of inclination.

To prove the converse let $(\widehat{U,V}) = 0$ and assume $(\widehat{U_1,V_1}) > 0$. Find finite dimensional $U_2 \subset U$ and $V_2 \subset V$ with $U = U_1 \oplus U_2$ and $V = V_1 \oplus V_2$. We have that $U_2 + V_2$ is finite dimensional and $(U_1 + V_1) \cap (U_2 + V_2) = \{0\}$. By our assumption and 1.4.2 the space $U_1 + V_1$ is closed. Hence $(U_1+V_1)+(U_2+V_2) = U + V$ is closed. According to 1.4.2 this contradicts $(\widehat{U,V}) = 0$.

(b) Now suppose that $U + V$ is not closed, hence $(\widehat{U,V}) = 0$. Fix $\epsilon \in \,]0,1[$. Choose sequences $\{\epsilon_i\}_{i=}^{\infty}$ and $\{\beta_i\}_{i=1}^{\infty}$ of numbers from $]0,1[$ which satisfy

$$\beta := \prod_{i=1}^{\infty} \beta_i > 0 \quad \text{and} \quad \sum_{i=1}^{\infty} \epsilon_i < \frac{\epsilon\beta^2}{1+\beta}$$

Since $(\widehat{U,V}) = 0$ there are elements $e_1 \in U$ and $g_1 \in V$ such that $\|e_1\| = 1$ and $\|e_1 - g_1\| < \epsilon_1$. By 1.4.4 there is a finite codimensional subspace U_1 in U such that $(\widehat{e_1,U_1}) > \beta_1$. Now (a) yields $(\widehat{U_1,V}) = 0$. Hence we find $e_2 \in U_1$ and $g_2 \in V$ with $\|e_2\| = 1$ and $\|e_2 - g_2\| < \epsilon_2$. Obviously, $(\widehat{e_1,e_2}) \geq (\widehat{e_1,U_1}) \geq \beta_1$. Put $L_{1,2} = \text{span}\{e_1,e_2\}$. Find a finite codimensional subspace $U_2 \subset U$ with $(\widehat{L_{1,2},U_2}) \geq \beta_2$. Again, (a) yields $(\widehat{U_2,V}) = 0$. Continuation of this procedure provides us with sequences $\{e_i\}_{i=1}^{\infty}$ of elements $e_i \in U$ and $\{g_i\}_{i=1}^{\infty}$ of elements $g_i \in V$ such that

$$\|e_i\| = 1, \quad (\widehat{L_{1,i},e_{i+1}}) \geq \beta_i \quad \text{and} \quad \|e_i - g_i\| < \epsilon_i, \quad i = 1,2,\dots,$$

where $L_{1,i} = \text{span}\{e_1,\dots,e_i\}$.

Put $U_0 = \text{closed span}\{e_i\}_{i=1}^{\infty}$ and $V_0 = \text{closed span}\{g_i\}_{i=1}^{\infty}$. Then $\{e_i\}_{i=1}^{\infty}$ is a basis of U_0 in view of 1.2.7 with $\gamma(\{e_i\}_{i=1}^{\infty}) \geq \beta$. Define $T : U_0 \to V_0$ by $T(\sum_{k=1}^{\infty} \alpha_k e_k) = \sum_{k=1}^{\infty} \alpha_k g_k$. In view of 1.2.9 we obtain, using the assumptions on ϵ_i and β_i,

$$\|T - id\| \leq \sum_{i=1}^{\infty} \frac{1+\beta}{\beta^2}\epsilon_i < \epsilon$$

This implies $\|T\| \leq 1 + \epsilon$. Moreover, T is invertible and

$$\|T^{-1}\| \leq \sum_{k=0}^{\infty} \|id - T\|^k \leq \frac{1}{1-\epsilon}$$

Hence $\max(\|T\|, \|T^{-1}\|) \leq 1+\epsilon/(1-\epsilon)$. Since ϵ was arbitrary we arrive at the contradiction that U and V are partially almost isometric. Thus $(\widehat{U,V}) > 0$ and $U + V$ is closed, □

The argument of the last part of the proof of 1.4.8 will be used again for the proof of the stability Theorem 2.7.2.

1.5 Projection Constants

If X is a complemented subspace of the Banach space E then it is natural to determine the "quality" of this complementation by introducing the *relative projection constant* (see [81] and [138])

$$\lambda(X, E) = \inf\{\|P\| \ : \ P : E \to X \text{ is a bounded projection}\} \ .$$

It is obvious that for $X \subset Y \subset E$ "monotonicity" holds:

$$\lambda(X, Y) \leq \lambda(X, E) \ .$$

We also define the *absolute projection constant* of X as a quantity which reflects the "worst" embedding of X into a Banach space Y:

$$\lambda(X) = \sup\{\lambda(X, Y) \ : \ Y \text{ a Banach space containing } X\} \ .$$

If there is a Banach space $Y \supset X$ in which X is non-complemented then we put $\lambda(X) = \infty$. Recall that $\lambda(X) < \infty$ whenever X is finite dimensional. Obviously, for every Banach space $Y \supset X$, we have

$$\lambda(X, Y) \leq \lambda(X) \ .$$

For a compact Hausdorff space K we denote by $C(K)$ the Banach space of all realvalued continuous functions on K endowed with the sup-norm.

Proposition 1.5.1 *Let E be a $C(K)$-space where K is a compact Hausdorff space and let X be a finite dimensional subspace of E. Then we have*

$$\lambda(X, E) = \lambda(X) \ .$$

Proof. We always have $\lambda(X, E) \leq \lambda(X)$.

To prove the opposite inequality let $E \supset X$ be a $C(K)$-space and consider a bounded projection $Q : E \to X$. Now let Y be any Banach space containing X. We claim that, for every $\epsilon > 0$, there is a linear operator $T : Y \to E$ with $\|T\| \leq 1 + \epsilon$ and $T|_X = id_X$. Indeed, using a suitable partition of unity, we find finitely many $f_1, \ldots, f_n \in E$ with

$$\max_k |\alpha_k| \leq \left\| \sum_{k=1}^n \alpha_k f_k \right\| \leq (1 + \epsilon) \max_k |\alpha_k|$$

for all choices of α_k such that $X \subset F := \text{span}\{f_1, \ldots, f_n\}$. Let $f_k^* \in E^*$ be linear functionals with $\|f_k^*\| = 1$ and

$$f_k^*(f_j) = \begin{cases} 1, & j = k \\ 0, & \text{else} \end{cases}$$

Such f_k^* exist by the Hahn-Banach theorem. Fix an arbitrary bounded projection $Q : E \to X$ and let $y_1^*, \ldots, y_n^* \in Y$ be such that $y_k^*|_X = f_k^*|_X$ and $||y_k^*|| = ||f_k^*|_X|| \le 1$ for all k. Put $Ty = \sum_{k=1}^n y_k^*(y)f_k$. This implies $Tx = x$ if $x \in X$ and $||Ty|| \le (1+\epsilon)||y||$ for all $y \in Y$. Define $P : Y \to X$ by $P = QT$. Then P is a projection onto X and we have $||P|| \le (1+\epsilon)||Q||$. We conclude $\lambda(X) \le (1+\epsilon)\lambda(X, E)$. Since ϵ was arbitrary we infer $\lambda(X) \le \lambda(X, E)$.   □

Kadec and Snobar [69] proved that for any n-dimensional Banach space X_n we have $\lambda(X_n) \le \sqrt{n}$, $n = 1, 2, \ldots$. König [73] showed that this inequality is strict for $n = 2, 3, \ldots$: $\lambda(X_n) < \sqrt{n}$.

Let us give the values of $\lambda(X_n)$ for some finite dimensional spaces ([37, 126, 138] Theorem 32.6):

Theorem 1.5.2 *For any positive integer n we have*

$$\lambda(l_p^n) \sim \sqrt{n} \quad if\ 1 \le p \le 2,$$
$$\lambda(l_p^n) \sim n^{1/p} \quad if\ 2 < p < \infty, \qquad and$$
$$\lambda(l_\infty^n) = 1.$$

Here l_p^n is the space $\mathbf{R}^n$ endowed with the norm

$$||(\alpha_1, \ldots, \alpha_n)|| = \begin{cases} (\sum_{k=1}^n |\alpha_k|^p)^{1/p} & \text{if } 1 \le p < \infty \\ \sup_k |\alpha_k| & \text{if } p = \infty \end{cases}$$

The only trivial part in the preceding theorem is $\lambda(l_\infty^n) = 1$. Indeed, let E be a Banach space with $E \supset l_\infty^n$. Consider the elements $e_1, \ldots, e_n \in l_\infty^n$ with $||\sum_{k=1}^n \alpha_k e_k|| = \max_k |\alpha_k|$ for all α_k and find, using the Hahn-Banach theorem, functionals $e_1^*, \ldots, e_n^* \in E^*$ with $||e_k^*|| = 1$ for all k and $e_k^*(e_j) = \begin{cases} 1, & j = k \\ 0, & \text{else} \end{cases}$. Put, for $e \in E$, $Pe = \sum_{k=1}^n e_k^*(e)e_k$. Then $P : E \to l_\infty^n$ is a projection with $||P|| = 1$.

l_∞^n is indeed the only finite dimensional Banach space with projection constant 1.

Theorem 1.5.3 *Let X be an n-dimensional Banach space with $\lambda(X) = 1$. Then X is isometrically isomorphic to l_∞^n.*

Proof. Let B_{X^*} be the closed unit ball of X^* and exB_{X^*} the set of extreme points of B_{X^*}. (a is an extreme point of the convex set A if $a = b = c$ whenever $b, c \in A$ and $0 < \lambda < 1$ such that $a = \lambda b + (1 - \lambda)c$.) Denote the w^*-closure of exB_{X^*} by K. Hence we have $K = -K$. Let

$$Y = \{f \ : \ f : K \to \mathbf{R} \text{ continuous and } f(x^*) = -f(-x^*) \text{ whenever } x^* \in K\}$$

Y is a closed subspace of $C(K)$. We may identify $x \in X$ with the function $\hat{x} \in Y$ where $\hat{x}(x^*) = x^*(x)$, $x^* \in K$. Then we have $||\hat{x}|| = ||x||$. Hence we can identify X with the subspace $\hat{X} = \{\hat{x} \ : \ x \in X\}$ of Y.

We claim that $\hat{X} = Y$. To prove the claim fix $x^* \in exB_{X^*}$ and put

$$F_{x^*} = \{\mu \in C(K)^* \ : \ ||\mu|| = 1 \text{ and } \mu(\hat{x}) = x^*(x) \text{ for all } x \in X\} \ .$$

Since x^* is an extreme point of B_{X^*} the set F_{x^*} is a face of the unit ball of $C(K)^*$, i.e. whenever $\nu_1, \nu_2 \in C(K)^*$ and $0 \leq \lambda \leq 1$ such that $||\nu_1||, ||\nu_2|| \leq 1$ and $\lambda\nu_1 + (1-\lambda)\nu_2 \in F_{x^*}$ then $\nu_1, \nu_2 \in F_{x^*}$. Moreover F_{x^*} is w^*-compact and convex. Hence by the Krein-Milman theorem F_{x^*} has an extreme point. The face property of F_{x^*} implies that every extreme point of F_{x^*} is an extreme point of the unit ball of $C(K)^*$. The latter extreme points are of the form $\pm\delta_k$ for some $k \in K$, where $\delta_k(f) = f(k)$, $f \in C(K)$. We conclude that $exF_{x^*} = \{\delta_{x^*}, -\delta_{-x^*}\}$ and

$$F_{x^*} = \{\lambda\delta_{x^*} - (1-\lambda)\delta_{-x^*} \ : \ 0 \leq \lambda \leq 1\} \ .$$

We have $\delta_{x^*}(y) = -\delta_{-x^*}(y)$ for all $y \in Y$ which implies $F_{x^*}|_Y = \{\delta_{x^*}|_Y\}$.

By assumption we find a linear $P : Y \to X$ with $P\hat{x} = x$ for all $x \in X$ and $||P|| = 1$. The adjoint operator P^* maps B_{X^*} onto a w^*-compact convex subset of B_{Y^*}, the closed unit ball of Y^*. We even have $P^*B_{X^*} = B_{Y^*}$. Indeed otherwise, by the separation theorem, we find $y \in Y$ with $||y|| > \sup_{x^* \in exB_{X^*}} |x^*(Py)|$. But for $x^* \in exB_{X^*}$ we have $x^* \circ P \in F_{x^*}|_Y$, i.e. $x^* \circ P = \delta_{x^*}|_Y$, and thus

$$||y|| = \sup_{k \in K} |y(k)| = \sup_{x^* \in exB_{X^*}} |(x^* \circ P)(y)|$$

since exB_{X^*} is dense in K. We arrive at a contradiction which shows that $P^*(B_{X^*}) = B_{Y^*}$. Therefore $P : Y \to X$ is an isometric isomorphism proving our claim.

For any $x_0^*, x_1^*, \ldots, x_m^* \in exB_{X^*}$ with $x_i^* \neq \pm x_j^*$ if $i \neq j$ the Tietze extension theorem yields a function $f \in C(K)$ with $f(\pm x_0^*) = \pm 1$ and $f(\pm x_j^*) = 0$ if $j > 0$. Let $y \in Y$ be the function with $y(x^*) = 2^{-1}f(x^*) - 2^{-1}f(-x^*)$, $x^* \in K$. Then $y(x_0^*) = 1$ and $y(x_j^*) = 0$ if $j > 0$. Since $Y \cong X$ was n-dimensional we conclude $exB_{X^*} = \{\pm x_1^*, \ldots, \pm x_n^*\}$ for some x_j^* and $Y = C(\{x_1^*, \ldots, x_n^*\}) = l_\infty^n$. $\qquad\square$

A more elaborate version of the preceding proof shows that an arbitrary Banach space X satisfies $\lambda(X) = 1$ if and only if X is isometrically isomorphic to a $C(K)$-space with an extremally disconnected compact Hausdorff space K (i.e. where the closure of any open subset of K is open again), see [71] and [34, 111].

A compactness argument involving families of projections onto finite dimensional subspaces yields the next proposition.

Proposition 1.5.4 *Let $E_1 \subset E_2 \subset \ldots$ be a sequence of finite dimensional subspaces of E with $\overline{\cup_{k=1}^{\infty} E_k} = E$. Then*

$$\lim_{k \to \infty} \lambda(E_1, E_k) = \lambda(E_1, E) .$$

Proof. By assumption we have

$$\lambda(E_1, E_2) \leq \lambda(E_1, E_3) \leq \ldots \leq \lambda(E_1, E) < \infty .$$

Hence $\lim_{n \to \infty} \lambda(E_1, E_n)$ exists and satisfies

$$\lim_{n \to \infty} \lambda(E_1, E_n) \leq \lambda(E_1, E) .$$

Conversely, take, for each n, a projection $P_n : E_n \to E_1$ with $||P_n|| \leq \lambda(E_1, E_n) + 1/n$. Hence $\sup_n ||P_n|| \leq \lambda(E_1, E) + 1$. Let

$$K = \{(\lambda(E_1, E) + 1)x \ : \ x \in E_1, \ ||x|| \leq 1\} .$$

Then K is compact since E_1 is finite dimensional. Let B_E be the closed unit ball of E and consider the element $\gamma_n := \{P_n e\}_{e \in B_E}$ of K^{B_E}. Since K^{B_E} is compact we find an accumulation point $\{x_e\}_{e \in B_E}$ of the sequence $\{\gamma_n\}_{n=1}^{\infty}$. Define, for $e \in E \setminus \{0\}$, $Pe = ||e||x_{e/||e||}$. Then $P : E \to E_1$ is a bounded projection with $||P|| \leq \limsup_{n \to \infty} ||P_n|| \leq \lim_{n \to \infty} \lambda(E_1, E_n)$. This implies $\lambda(E_1, E) \leq \lim_{n \to \infty} \lambda(E_1, E_n)$ and completes the proof. □

Definition 1.5.5 *For a complemented subspace $X \subset E$ the quantity*

$$l(X, E) = \lambda(X, E)/\lambda(X)$$

will be called the load *of X in E.*
If, for $\alpha \in [0, 1]$, $l(X, E) \leq \alpha$ then X is called α-unloaded in E.
If $l(X, E) = 1$ then X is called loaded.
A sequence of subspaces $X_1 \subset X_2 \subset \ldots$ in E is called unloaded *if $\lim_{k \to \infty} l(X_k, E) = 0$.*

Proposition 1.5.1 tells us that each subspace of a $C(K)$-space is loaded.

The notion of load will be used in 10.4. There we need some facts about the behaviour of $\lambda(X)$, $\lambda(X, E)$ and $l(X, E)$ under isomorphisms.

Proposition 1.5.6 *(a) Let $S : X \to Y$ be an isomorphism between the Banach spaces X and Y. Then we have*

$$\lambda(SX) \leq (||S|| \cdot ||S^{-1}||)\lambda(X) \quad and \quad \lambda(X) \leq (||S|| \cdot ||S^{-1}||)\lambda(SX)$$

(b) Let X be a complemented subspace of the Banach space E. Furthermore let F be another Banach space and assume that $T : E \to F$ is an (onto)-isomorphism. Then we have

$$\lambda(TX, F) \leq (||T|| \cdot ||T^{-1}||)\lambda(X, E) \quad and \quad l(TX, F) \leq (||T|| \cdot ||T^{-1}||)^2 l(X, E)$$

Proof. (a): We may assume without loss of generality $||S|| = 1$. Then we have $1 \leq ||S|| \cdot ||S^{-1}|| = ||S^{-1}||$.

We claim that, for any Banach space $E \supset X$, there is a Banach space $F \supset SX$ and an (onto-)isomorphism $T : E \to F$ with $T|_X = S$ and $||T|| = ||S||$, $||T^{-1}|| = ||S^{-1}||$.

Indeed, let $U = \{(e, Sx) \ : \ e \in E, x \in X\}$ be endowed with $||(e, Sx)|| = ||e|| + ||Sx||$. Consider the subspace V of U defined by $V = \{(x, -Sx) \ : \ x \in X\}$. Put $F = U/V$. Identify $y \in SX$ with $(0, y) + V$ in F. This identification is an isometry. Indeed, we have

$$||y|| \geq ||(0, y) + V|| = \inf_{x \in X} (||x|| + ||y - Sx||)$$
$$\geq \inf_{x \in X} (||x|| + ||y|| - ||S|| \cdot ||x||)$$
$$= ||y||$$

From now on we regard SX as a subspace of F. Define $T : E \to F$ by $Te = (e, 0) + V$, $e \in E$. Then, if $x \in X$, we have $Tx = (0, Sx) + V$ which corresponds to Sx. Moreover, for a general element $e \in E$ we obtain

$$||e|| \geq ||Te|| = \inf_{x \in X} (||e - x|| + ||Sx||)$$
$$\geq \inf_{x \in X} \left(\frac{1}{||S^{-1}||} ||e|| - \frac{1}{||S^{-1}||} ||x|| + ||Sx|| \right)$$
$$\geq \frac{1}{||S^{-1}||} ||e||$$

This proves the claim. The claim yields $\lambda(X) \leq ||S|| \cdot ||S^{-1}|| \lambda(SX)$. By exchanging X and SX we infer $\lambda(SX) \leq ||S|| \cdot ||S^{-1}|| \lambda(X)$.
(b) is an easy consequence of (a).   □

The argument of the proof of 1.5.6 (a) will be used again in the proof of Lemma 4.3.1.

1.6 The Projection Function of a Banach Space

For a given infinite dimensional Banach space E we define the following *projection function* for $t \in [1, \infty[$ [48]:

$$f_E(t) = \inf\{\lambda(X, E) \ : \ X \text{ a subspace of } E, \ \dim X < \infty, \ \lambda(X) \geq t\} \,.$$

The definition of $f_E(t)$ makes sense for every $t \geq 1$ since every infinite dimensional Banach space E contains a finite dimensional subspace X with $\lambda(X) \geq t$. Indeed, a famous theorem of Dvoretsky [24] states that, for every n and every $\epsilon > 0$, there is a subspace $X \subset E$ and an isomorphism $S : X \to l_2^n$ with $||S|| \cdot ||S^{-1}|| \leq 1 + \epsilon$. Hence we obtain

$$\sup\{\lambda(X) \ : \ X \text{ a subspace of } E, \ \dim X < \infty\} = \infty$$

in view of 1.5.2 and 1.5.6.

If $f_E(t) \sim t$, i.e. if $f_E(t)$ is asymptotically equivalent to t for $t \to \infty$, then E is said to have the *C-projection type*. If $f_E(t)$ is bounded, i.e. if $f_E(t) \sim 1$, then E is said to have the *L-projection type*.

In view of Proposition 1.5.1 every $C(K)$-space has the C-projection type. On the other hand, l_p, $1 \le p < \infty$, has the L-projection type. This follows from the fact that $\lambda(l_p^n, l_p) = 1$ but $\sup_n \lambda(l_p^n) = \infty$ according to 1.5.2.

A Banach space which has either C- or L-projection type will be said to have *standard projection type*, otherwise it will be of *non-standard projection type*.

It is not clear whether a Banach space of non-standard projection type exists at all:

Problems 1.6.1 *Does every infinite dimensional separable Banach space have standard projection type?*

A Banach space with non-standard projection type cannot have a basis.

Proposition 1.6.2 *Let E be a Banach space with basis. Then E has standard projection type.*

Proof. Let $\{e_k\}_{k=1}^{\infty}$ be a basis of E and put $L_n = \operatorname{span}\{e_1, \ldots, e_n\}$. 1.2.3 implies that $\sup_n \lambda(L_n, E) < \infty$. So, if $\sup_n \lambda(L_n) = \infty$, then E has L-projection type.

Now assume that $c := \sup_n \lambda(L_n) < \infty$. Fix t and consider any finite dimensional subspace $X \subset E$ with $\lambda(X) > t$. In view of 1.5.6 we may assume that $X \subset L_n$ for suitable large n. Let Y be a $C(K)$-space with $Y \supset L_n$ (for example, take K to be the closed unit ball of L_n^* under the w^*-topology). With 1.5.6 we conclude

$$\lambda(X) = \lambda(X, Y) \le \lambda(X, L_n)\lambda(L_n, Y) \le \lambda(X, E)c \le c\lambda(X) \,.$$

Hence, in this case, E has C-projection type. $\qquad\qquad\square$

Proposition 1.6.2 is actually true under weaker assumptions (see the end of Sect. 5.1).

Pisier, [121], constructed a separable Banach space E in which every n-dimensional subspace X_n satisfies $\lambda(X_n, E) \sim \sqrt{n}$. In view of the result of Kadec and Snobar mentioned in Sect. 1.4 this space E has C-projection type.

2

Sequences in Normed Spaces

Here we study, among other things, completeness and minimality of sequences in Banach spaces. This includes a discussion of bases and finite dimensional Schauder decompositions. Moreover this gives rise to the introduction of uniformly convex and uniformly smooth Banach spaces.

2.1 Complete, Supercomplete and $\{c_k\}$-Complete Sequences

For a given sequence $\bar{e} = \{e_k\}_{k=1}^{\infty}$ in a normed space E we sometimes write $L_n(\bar{e}) = \text{span } \{e_k\}_{k=1}^{n}$. Two sequences $\bar{e}$ and $\bar{g}$ in E will be called *similar* if $L_n(\bar{e}) = L_n(\bar{g})$, $n = 1, 2, \ldots$

As an example we mention the similarity of two sequences of polynomials $\{p_n(t)\}_{n=0}^{\infty}$ and $\{q_n(t)\}_{n=0}^{\infty}$ with degree $p_n = \text{degree } q_n = n$ in $C[0,1]$, say. This simple fact will be used in Part II in connection with various Müntz sequences.

Recall (see 1.2), a sequence $\bar{e}$ in E is complete if $\text{span}(\bar{e})$ is dense in E.

Definition 2.1.1 *The sequence $\bar{e}$ in E is called*
supercomplete *if $\{e_k\}_{k=n}^{\infty}$ is complete in E for $n = 1, 2, \ldots$,*
hypercomplete *if each subsequence of $\bar{e}$ is complete in E.*

Using the Weierstraß theorem we see that $\{t^k\}_{k=0}^{\infty}$ in $C[0,1]$ is supercomplete but not hypercomplete since $\{t^{2^j}\}_{j=0}^{\infty}$ is not complete in $C[0,1]$ (see 6.1.5). The first known examples of hypercomplete sequences were the sequences $\{t^{\lambda_k}\}_{k=0}^{\infty}$ where $\lambda_0 = 0$ and $\lambda_k \uparrow \lambda$ for some $\lambda > 0$. (The hypercompleteness is a consequence of the Müntz theorem 6.1.5).

Lateron it turned out that there exist hypercomplete sequences in any infinite dimensional separable Banach space. In particular, we have the following

Theorem 2.1.2 *For each complete linearly independent sequence $\bar{e} = \{e_k\}_{k=1}^{\infty}$ in an infinite dimensional separable Banach space E let*

$$g_n = \sum_{j=1}^{n} j^{-n} \frac{e_j}{||e_j||}.$$

Then $\bar{g} = \{g_n\}_{n=1}^{\infty}$ is a hypercomplete sequence which is similar to $\bar{e}$.

Proof. Consider $g_n = \sum_{j=1}^{n} j^{-n} e_j / ||e_j||$, $n = 1, 2, \ldots$. Clearly $\bar{g} = \{g_n\}_{n=1}^{\infty}$ is similar to $\bar{e}$. We claim that $\bar{g}$ is hypercomplete. To this end let $\{g_{n_k}\}_{k=1}^{\infty}$ be an arbitrary subsequence of $\bar{g}$. If $e^* \in E^*$ is such that $e^*(g_{n_k}) = 0$ for all k then we have $\sum_{j=1}^{n_k} j^{-n_k} e^*(e_j / ||e_j||) = 0$. Hence

$$\left| e^* \left(\frac{e_1}{||e_1||} \right) \right| = \left| -\sum_{j=2}^{n_k} \frac{1}{j^{n_k}} e^* \left(\frac{e_j}{||e_j||} \right) \right| \le ||e^*|| \sum_{j=2}^{n_k} \frac{1}{j^{n_k}} \le \frac{n_k - 1}{2^{n_k}} ||e^*||$$

for all k. This implies $e^*(e_1) = 0$. Similarly we obtain

$$\left| e^* \left(\frac{e_2}{||e_2||} \right) \right| = \left| -2^{n_k} \sum_{j=3}^{n_k} \frac{1}{j^{n_k}} e^* \left(\frac{e_j}{||e_j||} \right) \right| \le (n_k - 2) \left(\frac{2}{3} \right)^{n_k} ||e^*||$$

for all k and hence $e^*(e_2) = 0$. Continuing in this fashion we arrive at $e^*(e_k) = 0$ for all k and thus $e^* = 0$ since $\bar{e}$ is complete. The separation theorem implies that $\overline{\text{span}}\{g_{n_k}\}_{k=1}^{\infty} = E$ which shows that $\bar{g}$ is hypercomplete and proves the theorem. □

We note the following straighforward

Proposition 2.1.3 (a) Similar sequences $\bar{e}$ and $\bar{g}$ are either simultaneously complete or non-complete in E.

(b) If $\bar{e}$ is complete and $e_k \in \overline{\text{span}(\bar{g})}$ for each k then $\bar{g}$ is complete, too.

For a dual Banach space E^* the following notion is close to the concept of "completeness".

Definition 2.1.4 The sequence of functionals $e_k^* \in E^*$ is called total if, for each $x \in E$, the condition $e_k^*(x) = 0$, $k = 1, 2, \ldots$ implies $x = 0$.

We derive easily

Proposition 2.1.5 If the sequence $\bar{e}^* = \{e_k^*\}_{k=1}^{\infty} \subset E^*$ is complete in E^* then it is total.

Completeness of a sequence $\bar{e}$ does not imply essential geometric properties of $\bar{e}$ such as being a basis for E. This is shown by the following

Proposition 2.1.6 For any element x of an infinite dimensional separable Banach space E there is a hypercomplete sequence $\{g_n\}_{n=1}^{\infty}$ such that $\lim_{n \to \infty} g_n = x$.

Proof. Assume at first that $x \neq 0$. Let $\{e_k\}_{k=1}^{\infty}$ be a complete linearly independent sequence in E with $e_1 = x$. Put $g_n = ||x|| \sum_{j=1}^{n} j^{-n} e_j / ||e_j||$. According to 2.1.2, $\{g_n\}_{n=1}^{\infty}$ is hypercomplete. We have $||x - g_n|| \leq ||x||(n-1)/2^n$. Hence $\lim_{n \to \infty} g_n = x$.

If $x = 0$ then take any hypercomplete system $\{\tilde{g}_k\}_{k=1}^{\infty}$ and put

$$g_n = \frac{\tilde{g}_n}{n||\tilde{g}_n|| + 1} \, . \qquad \square$$

If $x \neq 0$ the sequence $\bar{g} = \{g_n\}_{n=1}^{\infty}$ in 2.1.6 satisfies

$$\lim_{n \to \infty} \left|\left| \frac{g_{n+1}}{||g_{n+1}||} - \frac{g_n}{||g_n||} \right|\right| = 0 \, .$$

Hence $\bar{g}$ cannot be a basis in view of 1.2.3 or 1.2.7 .

Completeness of a sequence $\bar{e} = \{e_k\}_{k=1}^{\infty}$ in E can be considered as an abstract version of the Weierstraß approximation theorem:

Given $x \in E$ and $\epsilon > 0$ there exists a "polynomial" $P_{x,\epsilon} = \sum_{k=1}^{m} a_k e_k$ such that $||x - P_{x,\epsilon}|| \leq \epsilon$.

Now consider a sequence $\bar{c}$ of positive numbers c_k such that, for any $x \in E$ and any $\epsilon > 0$, we find an approximation of x by $P_{x,\epsilon}$ as before where in addition $|a_k| \leq c_k ||x||$ for all k. Then $\bar{e}$ is called $\bar{c}$-*complete*. If we always find $P_{x,\epsilon}$ with $|a_k| \leq c_k$ for all k then $\bar{e}$ is called *super $\bar{c}$-complete*.

Theorem 2.1.7 *(See [55]) Each complete (supercomplete) sequence $\bar{e}$ in an infinite dimensional separable Banach space E is $\bar{c}$-complete (super $\bar{c}$-complete, resp.) for some sequence $\bar{c} = \bar{c}(\bar{e})$ of positive numbers.*

Proof. Let $\bar{e} = \{e_k\}_{k=1}^{\infty}$ be complete. We may assume that $\bar{e}$ is normalized.
a) At first we claim, there are positive numbers $\tilde{c}_k$ such that, for any n, any $x \in E$ and $\epsilon > 0$ there is $y = \sum_{k=1}^{m} \alpha_k e_k$ with $||x - y|| \leq \epsilon$, $m \geq n$ and $|\alpha_k| \leq \tilde{c}_k ||x||$ for $k = 1, 2, \ldots, n$.

Indeed, put $L_{k+1,\infty} = $ closed span$\{e_{k+1}, e_{k+2}, \ldots\}$ and $d_k = (e_k, \widehat{L_{k+1,\infty}})$, $k = 1, 2 \ldots$. Let $k_1 < k_2 < \ldots$ be the indices with $d_{k_j} \neq 0$. Then define $\tilde{c}_1 = \ldots = \tilde{c}_{k_1 - 1} = 0$ and $\tilde{c}_{k_1} = 1/d_{k_1}$. This clearly settles the case $n = k_1$.

Now assume that the claim is true for some $n = k_{m-1}$ with respect to $\tilde{c}_1, \ldots, \tilde{c}_{k_{m-1}}$.
Put $\tilde{c}_k = 0$ if $k_{m-1} + 1 \leq k \leq k_m - 1$ and

$$\tilde{c}_{k_m} = \frac{2 + \sum_{k=1}^{k_m - 1} \tilde{c}_k}{d_{k_m}} \, .$$

Then we obtain the case $n = k_m$.

Indeed, fix $x \neq 0$ and find, for each $\epsilon \in \,]0, 1[$, an element $y = \sum_k \alpha_k e_k$ with $||x - y|| \leq \epsilon ||x||$ and $|\alpha_k| \leq \tilde{c}_k ||x||$, $k = 1, \ldots, k_{m-1}$. Since $d_{k_{m-1}+1} = \ldots = d_{k_m - 1} = 0$ we may assume that $\alpha_{k_{m-1}+1} = \ldots = \alpha_{k_m - 1} = 0$. Then we have

$$2||x|| \geq ||y|| = \left\| \sum_{k=1}^{k_m-1} \alpha_k e_k + \alpha_{k_m} e_{k_m} + \sum_{k > k_m} \alpha_k e_k \right\|$$

$$\geq |\alpha_{k_m}| d_{k_m} - \sum_{k=1}^{k_m-1} \tilde{c}_k ||x||$$

which implies $|\alpha_{k_m}| \leq \tilde{c}_{k_m} ||x||$.

If $d_k = 0$ for all k then take $\tilde{c}_1 = \tilde{c}_2 = \ldots = 0$.

b) To show the general case take the sequence $\{\tilde{c}_k\}_{k=1}^{\infty}$ of a) and consider a dense sequence $\{x_k\}_{k=1}^{\infty}$ in E with $x_k \neq 0$ for all k. Find $y_1 = \sum_{k=1}^{n_1} \alpha_{1,k} e_k$ with $||x_1 - y_1|| \leq ||x_1||$ and $|\alpha_{1,1}| \leq \tilde{c}_1 ||x_1||$. Put

$$c_1 = 2\tilde{c}_1 \quad \text{and} \quad c_k = 2\max(|\alpha_{1,k}|, \tilde{c}_k), \quad k = 2, \ldots, n_1 .$$

Assume next that we have already $n_1 < n_2 < \ldots < n_{m-1}$ and $c_1, \ldots, c_{n_{m-1}}$. Then consider $y_m = \sum_{k=1}^{n_m} \alpha_{m,k} e_k$ with $n_m > n_{m-1}$, $||x_m - y_m|| \leq ||x_m||/m$ and $|\alpha_{m,k}| \leq \tilde{c}_k ||x_m||$, $k = 1, \ldots, n_{m-1}$. Put

$$c_k = 2\max(\tilde{c}_k, |\alpha_{m,k}|) \quad \text{if} \quad n_{m-1} + 1 \leq k \leq n_m .$$

Now, if $x \in E \setminus \{0\}$ is arbitrary and $\epsilon \in]0, 2[$ then find m with $||x - x_m|| \leq \epsilon ||x||/2$ and $1/m \leq \epsilon/4$. We obtain $||x_m|| \leq 2||x||$ and hence

$$||x - y_m|| \leq \epsilon ||x||, \quad y_m = \sum_{k=1}^{n_m} \alpha_{m,k} e_k \quad \text{and} \quad |\alpha_{m,k}| \leq \frac{1}{2} c_k ||x_m||$$

for all $k = 1, \ldots, n_m$. Since $||x_m|| \leq 2||x||$ we conclude $|\alpha_{m,k}| \leq c_k ||x||$ for all k. This proves the first part of 2.1.7.

c) Now let $\bar{e}$ be supercomplete. Let $c_{n,k} > 0$ be such that, for any x, any n and any $\epsilon > 0$, there is $y_n = \sum_{k \geq n} \alpha_{n,k} e_k$ with $||x - y_n|| \leq \epsilon$ and $|\alpha_{n,k}| \leq c_{n,k} ||x||$. The existence of the $c_{n,k}$ follows from b) and the fact that $\{e_k\}_{k=n}^{\infty}$ is complete. Fix $N \geq ||x||$ and put $y = N^{-1} \sum_{n=1}^{N} y_n$. Then we have $||x - y|| \leq \epsilon$ and $y = \sum_k N^{-1} \sum_{n=1}^{\min(k,N)} \alpha_{n,k} e_k$. We obtain

$$\left| \frac{1}{N} \sum_{n=1}^{\min(k,N)} \alpha_{n,k} \right| \leq \frac{1}{N} \sum_{n=1}^{k} c_{n,k} ||x|| \leq \sum_{n=1}^{k} c_{n,k} .$$

Put $c_k = \sum_{n=1}^{k} c_{n,k}$. This shows that $\bar{e}$ is super $\bar{c}$-complete. $\qquad \square$

In 6.4 we give estimates of the possible numbers $\bar{c}$ for the system $\{t^k\}_{k=0}^{\infty}$ in $C[0, 1]$.

We finish this subsection with the follwing simple but useful observation. A sequence $\{e_k\}_{k=1}^{\infty}$ is called *repeating* if, for each k, there is a subsequence $\{e_{n_j}\}_{j=1}^{\infty}$ such that $\lim_{j \to \infty} e_{n_j} = e_k$.

As a direct consequence of the definitions we obtain

Proposition 2.1.8 *Any complete repeating sequence is supercomplete.*

We leave it to the reader to construct, with Proposition 2.1.8, examples of supercomplete sequences.

2.2 Minimal and Uniformly Minimal Sequences

Let $\bar{e}$ be a sequence of non-zero elements in a Banach space E and put

$$L^{(k)} = \text{ closed span } \{e_1, \ldots, e_{k-1}, e_{k+1}, e_{k+2}, \ldots\} \,.$$

Recall, we defined $L_{i,j} = \text{span}\{e_i, e_{i+1}, \ldots, e_j\}$ and $L_{i,\infty} = \text{closed span}\{e_i, e_{i+1}, \ldots\}$.

Definition 2.2.1 $\bar{e}$ *is called* $\bar{\delta}$*-minimal if* $(e_k, \widehat{L^{(k)}}) \geq \delta_k$, $k = 1, 2, \ldots$, *for a sequence* $\bar{\delta} = \{\delta_k\}_{k=1}^{\infty}$ *of non-negative numbers.*

$\bar{e}$ is called minimal *if it is* $\bar{\delta}$*-minimal for some* $\bar{\delta} = \{\delta_k\}_{k=1}^{\infty}$ *of strictly positive numbers* δ_k.

If, in the preceding definition, $\delta_k \geq \delta > 0$ for all k and some $\delta > 0$ then $\bar{e}$ is called uniformly minimal *or δ-minimal (see 1.2.8). A 1-minimal system sometimes is called* Auerbach system.

$\bar{e}$ is called separated *if* $\inf_{j \neq k} (\widehat{e_j, e_k}) > 0$.

Finally, $\bar{e}$ is called closing *if*

$$\lim_{k \to \infty} (\widehat{e_k, e_{k+1}}) = 0 \,.$$

Obviously, we have

("uniformly minimal" $\Rightarrow$ "separated") and ("closing" $\Rightarrow$ "non-separated").

As a direct consequence of the definition we obtain that

$\bar{e}$ is minimal if and only if, for any k, $L^{(k)} \underset{\neq}{\subset} \text{closed span}(\bar{e})$

which explains the name.

Example. Consider $\bar{e} = \{t^{k^2}\}_{k=1}^{\infty}$ in $C[0,1]$. It will be shown in 6.1.5 and 9.2.2 that $\bar{e}$ is minimal but not uniformly minimal. On the other hand $\{t^k\}_{k=1}^{\infty}$ is not minimal in $C[0,1]$ although it is linearly independent. This is an easy consequence of the Weierstraß theorem. $\{t^k\}_{k=1}^{\infty}$ is even ω-linearly independent. (A sequence $\bar{e} = \{e_k\}_{k=1}^{\infty}$ is called ω-*linearly independent* if $\sum_{k=1}^{\infty} \alpha_k e_k = 0$ always implies $\alpha_1 = \alpha_2 = \ldots = 0$.)

Proposition 2.2.2 *Let $\bar{e} = \{e_k\}_{k=1}^{\infty}$ be a sequence in E with $e_k \neq 0$ for all k and assume that $\bar{\delta} = \{\delta_k\}_{k=1}^{\infty}$ is a sequence of positive numbers.*
(a) Then the following are equivalent.
   (i) $\bar{e} = \{e_k\}_{k=1}^{\infty}$ is $\bar{\delta}$-minimal.

(ii) For any $x = \sum_k \alpha_k e_k$ (finite sum or convergent series) we have

$$|\alpha_k| \leq \frac{||x||}{\delta_k ||e_k||}, \quad k = 1, 2, \ldots$$

(b) Let $\bar{\tau} = \{\tau_k\}_{k=1}^{\infty}$ be such that

$$(e_k, \widehat{L_{n+1,\infty}}) \geq \tau_n, n = 1, 2, \ldots, \text{ and } \tau_1 = \delta_1, \quad \delta_n \leq \tau_n \prod_{j=1}^{n-1} \frac{\tau_j}{1 + \tau_j}, \quad n = 2, 3, \ldots$$

Then $\bar{\tau}$ is $\bar{\delta}$-minimal.

Proof. (a): Let $Q : \mathrm{span}(\bar{e}) \to \mathrm{span}\{e_k\}$ be the projection with

$$Qe_j = \begin{cases} e_k & \text{if } j = k \\ 0 & \text{else} \end{cases}$$

Then, according to 1.2, we have $(e_k, \widehat{L^{(k)}}) = ||Q||^{-1}$. From this we infer immediately Proposition 2.2.2 (a).

(b): By assumption we obtain

$$\frac{\tau_{n-1}}{1 + \tau_{n-1}} \left\| \sum_{k \geq n} \alpha_k e_k \right\| \leq \left\| \sum_{k \geq n-1} \alpha_k e_k \right\| \quad \text{for all } \alpha_k \text{ and } n \geq 2 .$$

Using induction we see that this implies

$$\prod_{j=1}^{n-1} \frac{\tau_j}{1 + \tau_j} \left\| \sum_{k \geq n} \alpha_k e_k \right\| \leq \left\| \sum_{k \geq 1} \alpha_k e_k \right\| \quad \text{for all } \alpha_k \text{ and } n \geq 2 .$$

Hence

$$\left\| e_n + \sum_{j \neq n} \alpha_j e_j \right\| \geq \left\| e_n + \sum_{j > n} \alpha_j e_j \right\| \prod_{j=1}^{n-1} \frac{\tau_j}{1 + \tau_j} \geq \tau_n \prod_{j=1}^{n-1} \frac{\tau_j}{1 + \tau_j} \geq \delta_n .$$

We obtain $(e_n, \widehat{L^{(n)}})$ which proves (b).   □

Let $\bar{e}$ be a given sequence. Then $L_{1,n} = \mathrm{span}\{e_1, \ldots, e_n\}$ will be called the *n'th natural projection subspace* (or *Fourier projection subspace*).

$\nu_n(\bar{e}) = \nu_n = (L_{1,n}, \widehat{L_{n+1,\infty}})$ is the *n'th natural inclination number* and $\pi_n(\bar{e}) = \pi_n = 1/\nu_n$ the *n'th natural projection number* ([42], for the definition of $L_{i,j}$ see 1.2.2).

In the preceding proof we already used the following fact for a minimal sequence $\bar{e}$:

Let $P_n : \mathrm{span}(\bar{e}) \to L_{1,n}$ be the projection with $P_n(e_j) = 0$ for $j = n+1, n+2, \ldots$. Then we have $\pi_n(\bar{e}) = ||P_n||$.

We easily obtain the following

Corollary 2.2.3 *For a $\bar{\delta}$-minimal system $\bar{e}$ we have*

$$\nu_n(\bar{e}) \geq \frac{\min\{\delta_k\}_{k=1}^n}{n}$$

In particular, if $\bar{e}$ is δ-minimal for some $\delta > 0$ then

$$\nu_n(\bar{e}) \geq \frac{\delta}{n} \quad and \quad \pi_n(\bar{e}) \leq \frac{n}{\delta}$$

Proof. Let $P_n : \text{span}(\bar{e}) \to L_{1,n}$ be the projection with $P_n e_j = 0$ if $j > n$. 2.2.2 implies

$$||P_n|| \leq \frac{1}{\delta_1} + \ldots + \frac{1}{\delta_n} \leq \frac{n}{\min\{\delta_k\}_{k=1}^n}$$

Hence

$$\nu_n(\bar{e}) = (L_{1,n}, \widehat{L_{n+1,\infty}}) = \frac{1}{||P_n||} \geq \frac{\min\{\delta_k\}_{k=1}^n}{n} \qquad \square$$

So, for a 1-minimal system we have $\nu_n(\bar{e}) \geq 1/n$ and $\pi_n(\bar{e}) \leq n$. Sometimes we obtain better estimates.

For example, consider the trigonometric system in $E = C[0,1]$ where $e_0 = 1$, $e_{2k} = \cos(2\pi kt)$ and $e_{2k+1} = \sin(2\pi kt)$, $k = 1, 2, \ldots$. Then $\bar{e} = \{e_k\}_{k=0}^\infty$ is 1-minimal since for every $f = \sum_k \alpha_k e_k$ in $C[0,1]$ the Fourier coefficients satisfy $|\alpha_k| \leq ||f||$. It is known that $c_1 \log n \leq \pi_n(\bar{e}) \leq c_2 \log n$ for some absolute constants $c_1 > 0$ and $c_2 > 0$, [12].

Proposition 2.2.4 *The sequence $\bar{e} = \{e_k\}_{k=1}^\infty$ is minimal if and only if there exists a sequence $\bar{e}^* = \{e_k^*\}_{k=1}^\infty \subset E^*$ of functionals such that*

$$e_i^*(e_j) = \begin{cases} 1 & if \ i = j \\ 0 & if \ i \neq j \end{cases}$$

In this case $\bar{e}$ is $\bar{\delta}$-minimal with $\delta_k = (||e_k|| \cdot ||e_k^*||)^{-1}$.

Proof. Let $\bar{e}$ be $\bar{\delta}$-minimal for some $\bar{\delta} = \{\delta_k\}_{k=1}^\infty$ with positive δ_k. Then $\bar{e}$ must be linearly independent. Fix k and define the linear functional e_k^* on $\text{span}(\bar{e})$ by $e_k^*(\sum \alpha_j e_j) = \alpha_k$. Then, according to 2.2.2, $||e_k^*|| \leq (\delta_k ||e_k||)^{-1}$. Hence e_k^* can be extended to a bounded linear functional on E. We have $e_k^*(e_j) = 0$ if $k \neq j$.

Now assume that functionals $e_k^* \in E^*$ exist as indicated in the proposition. Put $x = \sum \alpha_j e_j$. We obtain

$$|\alpha_k| = e_k^*(x) \leq ||x|| \cdot ||e_k^*|| .$$

Hence, by 2.2.2, $\bar{e}$ is $\bar{\delta}$-minimal with $\delta_k = (||e_k|| \cdot ||e_k^*||)^{-1}$. $\qquad \square$

The sequence $\bar{e}^*$ is called *conjugate* to $\bar{e}$ and the sequence of pairs (e_j, e_j^*) is called *biorthogonal*. If $\bar{e}$ is complete in E then the conjugate sequence is uniquely determined.

In the preceding example where $\bar{e}$ is the trigonometric system in $C[0,1]$ let

$$e_0^*(f) = \int_0^1 f(t)dt \quad \text{and} \quad e_k^*(f) = 2\int_0^1 f(t)e_k(t)dt \quad \text{for all } f \in C[0,1] .$$

Then $\bar{e}^* = \{e_k^*\}_{k=1}^{\infty}$ is conjugate to $\bar{e}$. Here $\bar{e}$ is even complete in $C[0,1]$ and $\bar{e}^*$ is total.

Definition 2.2.5 *[52] If $\bar{e}$ is a complete minimal sequence in E such that the conjugate sequence $\bar{e}^*$ is total then $\bar{e}$ is called* Markushevich basis *or* M-basis.

In general, a Markushevich basis is not a basis. The preceding trigonometric system in $C[0,1]$ is a counterexample. Here $\bar{e}$ is not a basis since there are continuous functions whose Fourier series do not converge uniformly.

The following is due to Markushevich [102].

Proposition 2.2.6 *Each separable Banach space has a Markushevich basis.*

Proof. Let E be a separable Banach space. Fix $\{x_k\}_{k=1}^{\infty}$ in E such that $\text{span}\{x_k\}_{k=1}^{\infty}$ is dense in E and $x_k \neq 0$ for all k. Moreover, let $\{x_k^*\}_{k=1}^{\infty}$ be a total system in E^* which must exist since E is separable. (Here the unit ball B_{E^*} of E^* is w^*-metrizable and w^*-compact. Take, for example, $\{x_k^*\}_{k=1}^{\infty}$ to be w^*-dense in B_{E^*}.)

Define $\{e_k\}_{k=1}^{\infty}$ and $\{e_k^*\}_{k=1}^{\infty}$ by induction. Take $e_1 = x_1$. Let a be an index such that $x_a^*(x_1) \neq 0$ and put $e_1^* = x_a^*/x_a^*(x_1)$. Then let b be the smallest index with $x_b^* \notin \text{span}\{e_1^*\}$ and put $e_2^* = x_b^* - x_b^*(e_1)e_1^*$. Next let c be any index with $e_2^*(x_c) \neq 0$ and define $e_2 = (x_c - e_1^*(x_c)e_1)/e_2^*(x_c)$. Then we have

$$e_j^*(e_i) = \begin{cases} 1 & \text{if } i = j \\ 0 & \text{else} \end{cases} \quad \text{for } i, j = 1, 2 .$$

Now assume that we have already $e_1, \ldots, e_{2n}$ and $e_1^*, \ldots, e_{2n}^*$ with

$$e_j^*(e_i) = \begin{cases} 1 & \text{if } i = j \\ 0 & \text{else} \end{cases}$$

and $x_1, \ldots, x_n \in \text{span}\{e_j\}_{j=1}^{2n}$, $x_1^*, \ldots, x_n^* \in \text{span}\{e_j^*\}_{j=1}^{2n}$.

Let p be the smallest index with $x_p \notin \text{span}\{e_j\}_{j=1}^{2n}$ and put

$$e_{2n+1} = x_p - \sum_{j=1}^{2n} e_j^*(x_p)e_j .$$

Fix x_q^* such that $x_q^*(e_{2n+1}) \neq 0$ and define

$$e_{2n+1}^* = \frac{1}{x_q^*(e_{2n+1})} \left(x_q^* - \sum_{j=1}^{2n} x_q^*(e_j)e_j^* \right) .$$

Finally, let u be the smallest index with $x_u^* \notin \operatorname{span}\{e_j^*\}_{j=1}^{2n+1}$ and put

$$e_{2n+2}^* = x_u^* - \sum_{j=1}^{2n+1} x_u^*(e_j)e_j^* .$$

Fix x_v such that $e_{2n+2}^*(x_v) \neq 0$ and define

$$e_{2n+2} = \frac{1}{e_{2n+2}^*(x_v)} \left(x_v - \sum_{j=1}^{2n+1} e_j^*(x_v)e_j \right) .$$

Then we obtain

$$e_j^*(e_i) = \begin{cases} 1 & \text{if } i = j \\ 0 & \text{else} \end{cases} \quad \text{for } i, j = 1, \ldots, 2n+2$$

and $x_1, \ldots, x_{n+1} \in \operatorname{span}\{e_j\}_{j=1}^{2n+2}$, $x_1^*, \ldots, x_{n+1}^* \in \operatorname{span}\{e_j^*\}_{j=1}^{2n+2}$.

Thus, by construction, in view of 2.2.4, $\{e_j\}_{j=1}^{\infty}$ is a Markushevich basis of E. □

2.2.6 can be extended. To this end we need the following

Definition 2.2.7 *A Markushevich basis* $\bar{e} = \{e_k\}_{k=1}^{\infty}$ *with conjugate functionals* e_k^* *is called* hereditarily complete *or* strong *if, for each* $x \in E$,

$$x \in \overline{\operatorname{span}}\{ e_k \, : \, e_k^*(x) \neq 0 \} .$$

One can prove the following stronger version of Proposition 2.2.6

Theorem 2.2.8 *(See [50, 137]) Each separable Banach space has a strong Markushevich basis. If the Banach space is infinite dimensional then it also has a Markushevich basis which is not strong.*

An important case of a uniformly minimal sequence $\bar{e}$ is an *Auerbach sequence*, i.e. a 1-minimal sequence.

Theorem 2.2.9 *(See [3]) In each n-dimensional Banach space there exists an Auerbach system.*

Proof. Let $x_1, \ldots, x_n$ be a basis of the n-dimensional space E. For any set of elements $y_j = \sum_{k=1}^{n} \alpha_{j,k} x_k$ with $\|y_j\| \leq 1$, $j = 1, \ldots, n$, let $\Delta(y_1, \ldots, y_n)$ be the determinant of the matrix with the entries $\alpha_{j,k}$, $j, k = 1, \ldots, n$. Since the closed unit ball B_E of E is compact Δ attains its maximum at some n-tuple $(e_1, \ldots, e_n)$ where $e_j \in B_E$. The multilinearity of Δ ensures that $\|e_1\| = \ldots = \|e_n\| = 1$. Define

$$e_k^*(x) = \Delta(e_1, \ldots, e_{k-1}, x, e_{k+1}, \ldots, e_n)/\Delta(e_1, \ldots, e_n) \, .$$

Then we obtain $e_k^*(e_j) = \begin{cases} 1 & \text{if } k = j \\ 0 & \text{else} \end{cases}$ for all k and j and $||e_k^*|| = 1$ for all k.

2.2.2 shows that $\{e_k\}_{k=1}^n$ is an Auerbach system. □

It is completely unknown whether the infinite dimensional version of Theorem 2.2.9 is true. However one can show that every separable Banach space, for any $\epsilon > 0$, has a Markushevich basis $\{e_k\}_{k=1}^\infty$ with conjugate system $\{e_k^*\}_{k=1}^\infty$ such that $||e_k|| \cdot ||e_k^*|| \leq 1 + \epsilon$ for all k (see [114]).

2.3 Bases

Let $\bar{e} = \{e_k\}_{k=1}^\infty$ be a sequence in E with $e_k \neq 0$ for all k. In 1.2 we introduced the index

$$\gamma(\bar{e}) = \inf\{(L_{1,i}, \widehat{L_{i+1,j}}) \; : \; i < j\} \, .$$

We call $\nu(\bar{e}) = 1/\gamma(\bar{e})$ the *characteristic* of the sequence $\bar{e}$ and put $\nu(\bar{e}) = \infty$ if $\gamma(\bar{e}) = 0$ [42].

If $\bar{e}$ is linearly independent then we can define the following projections $P_m : \operatorname{span}(\bar{e}) \to L_{1,m}$ with $P_m(\sum_k \alpha_k e_k) = \sum_{k=1}^m \alpha_k e_k$. The definitions imply that

$$\nu(\bar{e}) = \sup_m ||P_m||$$

(see the remark after 1.2.1). With 1.2.3 we obtain

Proposition 2.3.1 $\bar{e}$ *is a basis of closed* $\operatorname{span}(\bar{e})$ *if and only if* $\nu(\bar{e}) < \infty$. *In this case we have*

$$\left|\left|\sum_{k=1}^n \alpha_k e_k\right|\right| \leq \nu(\bar{e}) \left|\left|\sum_{k=1}^\infty \alpha_k e_k\right|\right|$$

for all n and all choices of α_k.

If $\bar{e}$ is a basis of closed $\operatorname{span}(\bar{e}) \subset E$ then $\bar{e}$ is called *basic sequence* in E.

If $\bar{e}$ is a basis of E and $\bar{e}^*$ is the sequence conjugate to $\bar{e}$ then 2.3.1 implies that $\bar{e}^*$ is a basis sequence in E^*.

The characteristic $\nu(\bar{e})$ of a basic sequence, for obvious reasons, is also called the *basis constant* of $\bar{e}$.

Moreover, let

$$\gamma(E) = \sup\{\gamma(\bar{e}) \; : \; \bar{e} \text{ a complete sequence of non-zero elements in } E\}$$

be the *index* of E and $\nu(E) = 1/\gamma(E)$ the *characteristic of E*.

In 1972 Per Enflo constructed a separable Banach space E such that $\nu(E) = \infty$, i.e. this E does not have a basis, [27]. This gave a negative solution to the famous "basis problem."

Theorem 2.3.2 *There exists a separable Banach space without basis.*

But the "popular" Banach spaces have bases and sometimes even bases with special features. One is the the following

Definition 2.3.3 *Let $\bar{u} = \{u_k\}_{k=1}^{\infty}$ be a sequence of pairwise distinct elements of a compact metric space K. A basis $\bar{e} = \{e_k\}_{k=1}^{\infty}$ is called* interpolating *at the nodes $\bar{u}$ if for any $f = \sum_{k=1}^{\infty} \alpha_k e_k \in C(K)$ the n'th partial sum interpolates f at $u_1, \ldots, u_n$, i.e. if*

$$f(u_j) = \left(\sum_{k=1}^{n} \alpha_k e_k \right)(u_j), \quad j = 1, \ldots, n, \quad n = 1, 2, \ldots$$

It is easy to see that in this case necessarily $\bar{u}$ is dense in K. Indeed, by definition, for any $m > n$, with $f = e_m$, we obtain $e_m(u_n) = 0$. We claim, $e_m(u_m) \neq 0$ for all m. In fact, otherwise let $m_0 = \inf\{m : e_m(u_m) = 0\}$. Find $g \in C(K)$ with $g(u_m) = 0$ for $m < m_0$ and $g(m_0) = 1$, say $g = \sum_{k=1}^{\infty} \beta_k e_k$. We obtain $0 = g(u_1) = \beta_1 e_1(u_1)$ which implies $\beta_1 = 0$. Continuation yields $\beta_1 = \beta_2 = \ldots = \beta_{m_0-1} = 0$ and $1 = g(u_{m_0}) = \beta_{m_0} e_{m_0}(u_{m_0})$. Hence $e_{m_0}(u_{m_0}) \neq 0$, a contradiction. Thus $e_m(u_j) = 0$ if $m > j$ and $e_m(u_m) \neq 0$ for all m.

If $\bar{u}$ were not dense in K we would find $f \in C(K)$, $f \neq 0$, with $f(u_n) = 0$ for all n. By the preceding we would get $f = \sum_{k=1}^{\infty} \alpha_k e_k$ with $\alpha_1 = \alpha_2 = \ldots = 0$, a contradiction.

On the other hand we have

Theorem 2.3.4 *[46, 97] If $\bar{u}$ is a dense set of nodes in the compact metric space K then there exists a monotone basis of $C(K)$ which interpolates at $\bar{u}$.*

Here a basis $\bar{e}$ is called *monotone* if $\gamma(\bar{e}) = 1$. By definition this means,

$$\left\| \sum_{k=1}^{n} \alpha_k e_k \right\| \leq \left\| \sum_{k=1}^{\infty} \alpha_k e_k \right\| \quad \text{for all } n \text{ and all scalars } \alpha_k \,.$$

The proof of 2.3.4 is long and involved. We want to illustrate 2.3.4 with an example. Take $K = [0, 1]$ and fix a dense sequence $\bar{u} = \{u_k\}_{k=1}^{\infty}$ in $[0, 1]$ such that $u_1 = 0$, $u_2 = 1$ and $u_k \neq u_j$ if $j \neq k$. Let P_n be the operator which assigns to each $f \in C[0, 1]$ the piecewise affine function $P_n(f)$ with nodes at $u_1, \ldots, u_n$ such that $P_n(f)(u_k) = f(u_k)$, $k = 1, \ldots, n$. It is clear that P_n is linear in f and that $\|P_n(f)\| \leq \|f\|$ for each $f \in C[0, 1]$. If $e_{k,n}$ is the piecewise affine function with nodes at $u_1, \ldots, u_n$ such that

$$e_{k,n}(u_j) = \begin{cases} 1, \, j = k \\ 0, \, j \neq k \end{cases}, \quad k = 1, \ldots, n \,,$$

then $P_n(e_{k,n}) = e_{k,n}$. We obtain that $E_n := P_n C[0, 1] = \text{span}\{e_{1,n}, \ldots, e_{n,n}\}$. Hence P_n is a projection onto E_n with $\|P_n\| = 1$. If we put $e_n = e_{n,n}$ for all n then $\bar{e} = \{e_k\}_{k=1}^{\infty}$ is a basic sequence with basis projections $\{P_n\}_{n=1}^{\infty}$ since

$$P_n e_m = \begin{cases} e_m & \text{if } m < n \\ 0 & \text{else} \end{cases}$$

(in view of 2.3.1). Clearly, $\lim_{n\to\infty} ||P_n(f) - f|| = 0$ for any $f \in C[0,1]$. Hence $\bar{e}$ is a monotone basis of $C[0,1]$ which interpolates at $\bar{u}$.

Theorem 2.3.4 states that a similar construction of a basis is possible over any compact metric space for any prefixed dense set of nodes.

2.4 Uniformly Convex and Uniformly Smooth Spaces

Here we mention some basic facts about uniformly convex and uniformly smooth Banach spaces. Then, in the next section, we discuss bases in these special Banach spaces.

Definition 2.4.1 *The following function on E is called the* modulus of convexity *of E*

$$\delta_E(\omega) = \inf\left\{ 1 - \frac{1}{2}||x+y|| \; : \; x,y \in E,\; ||x|| \leq 1, ||y|| \leq 1,\; ||x-y|| \geq \omega \right\},$$

$0 \leq \omega \leq 2$. E is called uniformly convex if $\delta_E(\omega) > 0$ for all $\omega > 0$.
    The modulus of smoothness *of E [79] is the function*

$$\sigma_E(\tau) = \sup\left\{ \frac{||x+y|| + ||x-y||}{2} - 1 \; : \; x,y \in E,\; ||y|| \leq 1,\; ||x|| \leq \tau \right\}, \tau > 0.$$

E is called uniformly smooth *if $\lim_{\tau \to 0} \sigma_E(\tau)/\tau = 0$.*

It can be shown that in the definition of δ_E the inf can be taken over all x, y with $||x|| = ||y|| = 1$ and $||x-y|| \geq \omega$ without changing δ_E. Similarly, in the definition of σ_E the sup can be taken over all x and y with $||x|| = 1$ and $||y|| = \tau$ (see [84], II p.60).

Classical examples of uniformly convex spaces are L_p and l_p for $1 < p < \infty$. Their modulus of convexity is

$$\delta_E(\omega) \sim \frac{p-1}{2}\left(\frac{\omega}{2}\right)^2 \text{ if } 1 < p < 2 \text{ and } \delta_E(\omega) \sim \frac{1}{p}\left(\frac{\omega}{2}\right)^p \text{ if } p \geq 2,$$

see [18].

The notions of uniform convexity and uniform smoothness are in a way dual to each other. The following proposition is due to Šmul'yan, [133].

Proposition 2.4.2 *A Banach space E is uniformly convex if and only if E^* is uniformly smooth.*

Proof. [79] We claim that

$$\sigma_{E^*}(\tau) = \sup\left\{\frac{1}{2}\tau\omega - \delta_E(\omega) \; : \; 0 \le \omega \le 2\right\} \quad \text{for all } \tau > 0$$

Indeed, we have

$$
\begin{aligned}
2\sigma_{E^*}(\tau) &= \sup\{\|x^* + \tau y^*\| + \|x^* - \tau y^*\| - 2 \; : \; x^*, y^* \in E^*, \\
&\qquad \|x^*\| \le 1, \|y^*\| \le 1\} \\
&= \sup\{x^*(x) + \tau y^*(x) + x^*(y) - \tau y^*(y) - 2 \; : \; x \in E, y \in E, \\
&\qquad \|x\| \le 1, \|y\| \le 1, x^* \in E^*, y^* \in E^*, \|x^*\| \le 1, \|y^*\| \le 1\} \\
&= \sup\{\|x + y\| + \tau\|x - y\| - 2 \; : \; x \in E, y \in E, \\
&\qquad \|x\| \le 1, \|y\| \le 1\} \\
&= \sup\{\|x + y\| + \tau\omega - 2 \; : \; x \in E, y \in E, \|x\| \le 1, \|y\| \le 1, \\
&\qquad \|x - y\| \ge \omega, 0 \le \omega \le 2\} \\
&= \sup\{\tau\omega - 2\delta_E(\omega) \; : \; 0 \le \omega \le 2\}
\end{aligned}
$$

If E is uniformly convex then we have $\delta_E(\omega) > 0$ for all ω. By definition, $\delta_E(\omega)$ is increasing. Fix $\epsilon > 0$ and consider τ with $0 < \tau < \delta_E(\epsilon)$. We obtain

$$\omega \le 2\epsilon + \frac{2\delta_E(\epsilon)}{\tau} \le 2\epsilon + \frac{2\delta_E(\omega)}{\tau} \quad \text{if } \epsilon \le \omega \le 2$$

and, clearly, $\omega \le 2\epsilon + 2\delta_E(\omega)/\tau$ if $0 \le \omega \le \epsilon$. We conclude, using the claim, that $\sigma_{E^*}(\tau)/\tau \le \epsilon$. Hence $\lim_{\tau \to 0} \sigma_{E^*}(\tau)/\tau = 0$ and E^* is uniformly smooth.

Conversely, if E^* is uniformly smooth then, for every $\omega \in \,]0, 2[$, there exists a $\tau > 0$ such that $\sigma_{E^*}(\tau) \le \tau\omega/4$. The claim yields that $\tau\omega/2 - \delta_E(\omega) \le \tau\omega/4$ i.e. $\delta_E(\omega) \ge \tau\omega/4$. Hence $\delta_E(\omega) > 0$ for all $\omega > 0$ and E is uniformly convex. □

Uniform smoothness and uniform convexity are properties which depend on the isometric character of a Banach space. They may get lost if we go over to isomorphic Banach spaces. However, we have the following fundamental result, due to Enflo, [26].

Theorem 2.4.3 *A Banach space E is isomorphic to a uniformly convex space if and only if E is isomorphic to a uniformly smooth space.*

A Banach space E which is isomorphic to a uniformly convex space is called *superreflexive*.

Proposition 2.4.4 *[106] Every uniformly convex Banach space is reflexive.*

Proof. Regard E as subspace of E^{**} in the canonical way. Fix a norm one element $x^{**} \in E^{**}$ and find a directed set $\{x_\alpha\}_{\alpha \in A}$ of elements in E with $w^* - \lim_\alpha x_\alpha = x^{**}$ and $\|x_\alpha\| \le 1$ for all α. Hence $\lim_{\alpha,\beta} \|x_\alpha + x_\beta\| = 2\|x^{**}\| = 2$. The uniform convexity of E implies $\lim_{\alpha,\beta} \|x_\alpha - x_\beta\| = 0$. This means that $\{x_\alpha\}_{\alpha \in A}$ converges in norm and $\lim_\alpha \|x_\alpha - x^{**}\| = 0$, i.e. $x^{**} \in E$. Hence $E = E^{**}$. □

Let E be a uniformly smooth Banach space. Then, using the w^*-density of the unit ball of E in the unit ball of E^{**} (if we regard E as subspace of E^{**}) we see that E^{**} is uniformly smooth, too. Hence E^* is uniformly convex by 2.4.2. We conclude with 2.4.4 that every uniformly smooth Banach space E is reflexive, too (since E^* is reflexive).

Finally, we note the well-known result of the existence and uniqueness of best approximation in uniformly convex spaces.

Proposition 2.4.5 *Let E be a uniformly convex space and $X \subset E$ a closed subspace. Then, for any element $e \in E$, there exists a unique element $x_0 \in X$ of best approximation, i.e.*

$$||e - x_0|| = \inf_{x \in X} ||e - x|| .$$

Proof. Assume that $e \in E \setminus X$. Find $x_n \in X$ with

$$||e - x_n|| \le \inf_{x \in X} ||e - x|| + \frac{1}{n} .$$

Hence $\{x_n\}_{n=1}^\infty$ is bounded. Since E, and therefore X, is reflexive we find a weak accumulation point $x_0 \in X$ of $\{x_n\}_{n=1}^\infty$. We clearly have $b := ||e - x_0|| = \inf_{x \in X} ||e - x|| > 0$. Now let $y_0 \in X$ be another element with $||e - y_0|| = \inf_{x \in X} ||e - x||$. If $x_0 \ne y_0$ then, in view of the uniform convexity, we obtain, with $\omega = ||e - x_0 - (e - y_0)||/b = ||x_0 - y_0||/b > 0$, that $\delta_E(\omega) > 0$. Hence $||e - (x_0 + y_0)/2|| < b$ which contradicts the fact that x_0 and y_0 were optimal already. Thus $x_0 = y_0$. $\qquad\square$

2.5 Bases in Uniformly Convex Spaces

Here we collect some classical theorems about bases in uniformly convex spaces.

The proof of the following theorem is based on the interplay between the positivity of $\delta_E(\omega)$ and the positivity of the angle index $\varphi_E(\bar{e})$ of 1.2.9. Recall,

$$\varphi_E(\bar{e}) = \inf \left\{ \left\| \frac{x}{||x||} - \frac{y}{||y||} \right\| : x \in \operatorname{span}\{e_i\}_{i=1}^m, \ x \ne 0, \right.$$

$$\left. y \in \operatorname{span}\{e_i\}_{i=m+1}^{m+n}, \ y \ne 0, n, m = 1, 2, \dots \right\}$$

If $\gamma(\bar{e})$ is the index of $\bar{e}$ then we easily obtain

$$\gamma(\bar{e}) \le \varphi_E(\bar{e}) \le 2\gamma(\bar{e}) .$$

Hence, by 2.3.1, $\bar{e}$ is a basic sequence if and only if $\varphi_E(\bar{e}) > 0$.

Theorem 2.5.1 *[40] Let E be a uniformly convex space and $\bar{e} = \{e_k\}_{k=1}^{\infty}$ a normalized basis of E. Then there exist exponents $1 < \tau \leq \rho < \infty$ and numbers $A, B > 0$ (depending on $\bar{e}$) such that, for any element $x = \sum_k \alpha_k e_k$, the following inequalities hold*

$$B \left(\sum_k |\alpha_k|^\rho \right)^{1/\rho} \leq ||x|| \leq A \left(\sum_k |\alpha_k|^\tau \right)^{1/\tau}.$$

Proof. We need to show the right-hand inequality only. Indeed, then by dualizing, involving the conjugate sequence, we see that the left-hand inequality holds for any normalized basis in a uniformly smooth space. An application of 2.4.2 finishes the proof.

To prove the right-hand inequality let $\lambda = 2(1 - \delta_E(\varphi_E(\bar{e})/2))$ and fix a number τ with $1 < \tau < \log 2 / \log \lambda$. Observe that we have $\varphi_E(\bar{e})/2 \leq 1$ by definition. Hence $\delta_E(\varphi_E(\bar{e})/2) \leq 1/2$ and $\lambda \geq 1$.

We claim that there exists $\eta = \eta(\tau, \lambda) \in]0, 1[$ such that, for $x, y \in E$ with $||x|| \leq 1$ and $||y|| \leq 1$ and $||x - y|| \geq \varphi_E(\bar{e})$, we have

$$||x + ty||^\tau < 1 + t^\tau \quad \text{whenever } |1 - t| \leq \eta.$$

Indeed, the definition of δ_E implies $||x + y|| \leq \lambda$. Moreover we have $\lambda^\tau < 2$ since $\lambda \geq 1$. This proves the claim for $t = 1$. Since $||x + ty||^\tau \leq (\lambda + |1 - t|)^\tau$ and $(\lambda + |1 - t|)^\tau$ and $1 + t^\tau$ are continuous in t we derive the claim for t close enough to 1.

To finish the proof of the theorem let $x = \sum_{i=1}^{m} \alpha_i e_i$. Put $A = 4/\eta$. Then we show $||x|| \leq A \left(\sum_{i=1}^{m} |\alpha_i|^\tau \right)^{1/\tau}$ by induction on m.

If $m = 1$ then the assertion is clear since $A \geq 1$.

Suppose now that, for some n, $||x|| \leq A \left(\sum_{i=1}^{m} |\alpha_i|^\tau \right)^{1/\tau}$ for all $m \leq n$ and all choices of α_i.

Then, finally, consider $x = \sum_{i=1}^{n+1} \alpha_i e_i$. We distinguish between two cases.
Case 1: There is $i_0 \leq n + 1$ such that $4^{-1}\eta ||x|| \leq |\alpha_{i_0}|$. Then we have

$$||x|| \leq \frac{4}{\eta} |\alpha_{i_0}| \leq \frac{4}{\eta} \left(\sum_{i=1}^{n+1} |\alpha_i|^\tau \right)^{1/\tau}$$

which proves step $n + 1$ of the induction in this case. So we are left with
Case 2: $|\alpha_i| \leq 4^{-1}\eta ||x||$ for all i. Put

$$x_j = \sum_{i=1}^{j} \alpha_i e_i \quad \text{and} \quad y_j = \sum_{i=j+1}^{n+1} \alpha_i e_i \quad \text{and} \quad x_0 = 0 = y_{n+1}$$

Then we obtain $x = x_j + y_j$, $j = 0, 1, \ldots, n + 1$. Moreover we have

$$||x_0|| - ||y_0|| < 0 < ||x_{n+1}|| - ||y_{n+1}||$$

(provided that $x \neq 0$) and

$$\big|\, ||x_{j+1}|| - ||x_j||\,\big| \leq \frac{\eta}{4}||x|| \quad \text{and} \quad \big|\, ||y_{j+1}|| - ||y_j||\,\big| \leq \frac{\eta}{4}||x||\,.$$

Let $r = \sup\{j\ :\ ||x_j|| - ||y_j|| < 0\} + 1$. Then

$$||x_{r-1}|| - ||y_{r-1}|| < 0 < ||x_r|| - ||y_r|| \leq ||x_{r-1}|| - ||y_{r-1}|| + \frac{\eta}{2}||x||\,.$$

Hence $0 < ||x_r|| - ||y_r|| \leq 2^{-1}\eta||x||$. In particular $||y_r|| \leq ||x_r||$ and $0 < r < n+1$ since $0 < \eta < 1$. We may assume without loss of generality $||x_r|| = 1$ (otherwise take $x/||x_r||$ instead of x). Then we have, with $t = ||y_r||$, $0 \leq 1-t \leq \eta$ since $||x|| = ||x_r + y_r|| \leq 2$. The claim yields, since $||x_r - ||y_r||^{-1}y_r|| \geq \varphi_E(\bar{e})$,

$$||x||^\tau = \left\| x_r + t\frac{y_r}{||y_r||} \right\|^\tau < 1 + t^\tau = ||x_r||^\tau + ||y_r||^\tau\,.$$

Including the induction hypothesis we obtain

$$||x|| \leq A \left(\sum_{i=1}^{r} |\alpha_i|^\tau + \sum_{i=r+1}^{n+1} |\alpha_i|^\tau \right)^{1/\tau}$$

which concludes the proof. $\qquad\qquad\qquad\qquad\qquad\qquad\qquad\qquad\square$

Since $\sum_k k^{-\tau} < \infty$ if $1 < \tau$ we obtain

Corollary 2.5.2 *Under the assumptions of the preceding theorem the harmonic series $\sum_k e_k/k$ converges in norm.*

In a Hilbert space H, which is of course uniformly convex, one can give a quantitative generalization of Parseval's theorem with a proof which is similar to that of 2.5.1. In fact one can show:

Theorem 2.5.3 *[51] Let $\bar{e}$ be a normalized basis of the Hilbert space H. Assume that there is $\eta \in [0,1[$ satisfying $C(\bar{e}) \geq \eta$ where $C(\bar{e})$ is the cos-index of 1.b. Then the inequalities of Theorem 2.5.1 hold with*

$$\tau = \frac{2}{1+\sqrt{\eta}}, \quad \rho = \frac{2}{1-\sqrt{\eta}}, \quad A = 1+5\sqrt{\eta}, \quad B = \frac{1-\sqrt{\eta}}{1+5\sqrt{\eta}}\,.$$

In particular, for an orthonormal basis in H we have $C(\bar{e}) = 0$. Hence in this case 2.5.3 yields the Parseval theorem.

A normalized basis in an arbitrary Banach space is called ρ-*Besselian* (or τ-*Hilbertian*, resp.) if it satisfies the left-hand side (or right-hand side, resp.) of Theorem 2.5.1.

It was shown by Szarek [134] that $C[0,1]$ does not have a basis which is ρ-Besselian for some ρ. Combining this with 2.5.1 we obtain

Theorem 2.5.4 *A sequence of continuous functions* $x_n(t)$ *on* $[0,1]$ *with*

$$0 < \inf_n \inf_t |x_n(t)| \le \sup_n \sup_t |x_n(t)| < \infty$$

cannot be simultaneously a basis in C *and in* L_p *for any* $p \in\ ]1, \infty[$.

Proof. Assume that $\{x_n\}_{n=1}^\infty$ is a basis in C as well as in L_p for some $p \in\ ]1, \infty[$. Let $||\cdot||_C$ be the sup-norm in C and $||\cdot||_{L_p}$ the norm in L_p. Then $\{x_n/||x_n||_C\}_{n=1}^\infty$ is a normalized basis in C and $\{x_n/||x_n||_{L_p}\}_{n=1}^\infty$ is a normalized basis in L_p. By our assumption we find a constant $b > 0$ with $b \le ||x_n||_{L_p}/||x_n||_C$ for all n. Since L_p is uniformly convex, according to 2.5.1, we find some $\rho > 1$ and a constant $B > 0$ with

$$Bb\left(\sum_k |\alpha_k|^\rho\right)^{1/\rho} \le B\left(\sum_k\left(|\alpha_k|\frac{||x_k||_{L_p}}{||x_k||_C}\right)^\rho\right)^{1/\rho} \le \left|\left|\sum_k \alpha_k \frac{x_k}{||x_k||_C}\right|\right|_{L_p}$$

$$\le \left|\left|\sum_k \alpha_k \frac{x_k}{||x_k||_C}\right|\right|_C$$

for all choices of α_k. Hence $\{x_n/||x_n||_C\}_{n=1}^\infty$ is a ρ-Besselian basis in C which is impossible. □

From 2.3.3 we obtain the result, due to Olevskii [113], that a uniformly bounded sequence of continuous functions f_n on $[0,1]$ which is orthonormal in L_2 cannot be a basis in $C[0,1]$. This applies in particular to the trigonometric system (see 2.2). As a corollary we obtain the classical result that there are continuous functions on $[0,1]$ whose Fourier series does not converge uniformly.

Definition 2.5.5 *A basis* $\bar{e} = \{e_k\}_{k=1}^\infty$ *in a Banach space* E *is called unconditional if, for any permutation* π *of the indices, the series* $\sum_k \alpha_{\pi(k)} e_{\pi(k)}$ *converges whenever* $\sum_k \alpha_k e_k$ *converges.*

Theorem 2.5.6 *[66] Let* $\{e_k\}_{k=1}^\infty$ *be a normalized unconditional basis in a uniformly convex space* E. *Then, for each* $x = \sum_k \alpha_k e_k \in E$, *the series* $\sum_k \delta_E(|\alpha_k|)$ *converges.*

Proof. [83] Assume that for some $x = \sum_{k=1}^\infty \alpha_k e_k \in E$ we have $\sum_{k=1}^\infty \delta_E(|\alpha_k|) = \infty$. Choose signs $\epsilon_k \in \{1, -1\}$ and put $S_n = \sum_{k=1}^n \epsilon_k \alpha_k e_k$ such that

$$||S_{n-1} + \epsilon_n \alpha_n e_n|| \ge ||S_{n-1} - \epsilon_n \alpha_n e_n|| \quad \text{for all } n.$$

By assumption, $S := \lim_n S_n$ exists in E. We may assume without loss of generality $||S|| = 3/2$ such that $1 \le ||S_n|| \le 2$ for large enough n. Put

$$y_n = \frac{S_n}{||S_n||} \quad \text{and} \quad z_n = \frac{S_{n-1} - \epsilon_n \alpha_n e_n}{||S_n||}.$$

Hence $||y_n|| = 1$, $||z_n|| \leq 1$ and $||y_n - z_n|| = 2|\alpha_n|/||S_n||$. This implies, since δ_E is increasing,

$$\delta_E(|\alpha_n|) \leq \delta_E\left(\frac{2|\alpha_n|}{||S_n||}\right) \leq 1 - \frac{||y_n + z_n||}{2} = 1 - \frac{||S_{n-1}||}{||S_n||} \leq ||S_n|| - ||S_{n-1}||$$

for large enough n. We obtain

$$\sum_{k=n_0}^{n} \delta_E(|\alpha_k|) \leq ||S_n|| - ||S_{n_0}||$$

if n_0 is large enough. Now $\sum_{k=n_0}^{\infty} \delta_E(|\alpha_k|) = \infty$ contradicts the fact that $\lim_n ||S_n||$ exists. Hence the theorem follows. $\qquad\square$

In [16] the result of 2.5.6 was extended to normalized monotone bases. In [23] and, independently, in [119] it was shown that, in case of $E = L_p$ or $E = l_p$ for $1 < p < \infty$, monotone bases are always unconditional. We have

Theorem 2.5.7 *Each monotone basis in L_p or l_p for $1 < p < \infty$ is unconditional.*

2.6 Bases of Subspaces

Let $\{E_k\}_{k=1}^{\infty}$ be a sequence of closed subspaces of a Banach space E. $\{E_k\}_{k=1}^{\infty}$ will be called *basis of subspaces* for E if each $x \in E$ has a unique representation as

$$x = \sum_k e_k, \quad \text{where } e_k \in E_k, \quad k = 1, 2, \ldots .$$

Then we sometimes write $E = \sum_k \oplus E_k$. If all E_k are finite dimensional we also call $\{E_k\}_{k=1}^{\infty}$ a *finite dimensional Schauder decomposition (FDD)*, see Definition 5.1.1.

Repeating the argument of the proof of 1.2.3 we obtain

Proposition 2.6.1 *Let $\{E_k\}_{k=1}^{\infty}$ be a sequence of closed subspaces of E. Then the following are equivalent*

(i) $\{E_k\}_{k=1}^{\infty}$ *is a basis of subspaces of E*

(ii)   *There is some $\delta > 0$ such that*

$$(E_1 + E_2 + \ldots + E_n, \widehat{E_{n+1} + E_{n+2} + \ldots + E_{n+m}}) \geq \delta$$

whenever m and n are positive integers.

(iii) *There is a constant $\delta > 0$ such that, for all choices of $e_k \in E_k$ and all indices m and n,*

$$\delta \left\|\sum_{k=1}^{n} e_k\right\| \leq \left\|\sum_{k=1}^{m+n} e_k\right\|$$

Combining 1.2.3 with 2.6.1 we obtain

Theorem 2.6.2 *Let* $\{E_k\}_{k=1}^{\infty}$ *be a basis of subspaces of* E *such that all* E_k *are finite dimensional. Moreover, let* $\bar{e}_k = \{e_{k,j}\}_{j=1}^{m_k}$ *be bases of* E_k *with* $\inf_k \gamma(\bar{e}_k) > 0$. *Then*

$$e_{1,1}, \dots, e_{1,m_1}, e_{2,1}, \dots, e_{2,m_2}, e_{3,1}, \dots$$

is a basis of E.

Let E_k, $k = 1, 2, \dots$, be a sequence of Banach spaces. For $1 \le p < \infty$ put

$$\left(\sum_{k=1}^{\infty} \oplus E_k \right)_{(l_p)} = \left\{ \{e_k\}_{k=1}^{\infty} \ : \ e_k \in E_k \text{ for all } k, \sum_{k=1}^{\infty} ||e_k||^p < \infty \right\}$$

and consider the norm

$$||\{e_k\}_{k=1}^{\infty}|| = \left(\sum_{k=1}^{\infty} ||e_k||^p \right)^{1/p}.$$

Moreover we define

$$\left(\sum_{k=1}^{\infty} \oplus E_k \right)_{(l_\infty)} = \{ \{e_k\}_{k=1}^{\infty} \ : \ e_k \in E_k \text{ for all } k, \sup_k ||e_k|| < \infty \}$$

endowed with the norm $||\{e_k\}_{k=1}^{\infty}|| = \sup_k ||e_k||$ and

$$\left(\sum_{k=1}^{\infty} \oplus E_k \right)_{(c_0)} = \left\{ \{e_k\}_{k=1}^{\infty} \in \left(\sum_{k=1}^{\infty} \oplus E_k \right)_{(l_\infty)} \ : \ \lim_{k \to \infty} ||e_k|| = 0 \right\}.$$

Clearly, if $\sup_k \dim E_k < \infty$ then $(\sum_k \oplus E_k)_{(l_p)}$ is isomorphic to l_p and $(\sum_k \oplus E_k)_{(c_0)}$ is isomorphic to c_0. Similarly we define $(\sum_{k=1}^{n} \oplus E_k)_{(l_p)}$ for any positive integer n.

2.7 Stability of Sequences

We conclude Chap. 2 with some stability results. At first we introduce

Definition 2.7.1 *Let* $\bar{e} = \{e_k\}_{k=1}^{\infty}$ *and* $\bar{g} = \{g_k\}_{k=1}^{\infty}$ *be sequences in the Banach spaces* E *and* F, *resp. Then, for some* $a \ge 1$, $\bar{e}$ *and* $\bar{g}$ *are called* a-*equivalent if there is an isomorphism* T *from closed* $\mathrm{span}(\bar{e})$ *onto closed* $\mathrm{span}(\bar{g})$ *such that* $Te_k = g_k$ *for* $k = 1, 2, \dots$ *with* $||T|| \cdot ||T^{-1}|| \le a$.
  If $a = 1$ *then* $\bar{e}$ *and* $\bar{g}$ *are called* isometrically equivalent.

Then we prove a stability result for basic and minimal sequences.

Theorem 2.7.2 *[75] Let $\bar{e} = \{e_k\}_{k=1}^\infty$ be a $\bar{\delta}$-minimal sequence in E where $\bar{\delta} = \{\delta_k\}_{k=1}^\infty$ has positive δ_k.*
(a) For some $a > 1$ let $\bar{\epsilon} = \{\epsilon_k\}_{k=1}^\infty$ satisfy

$$\sum_{k=1}^\infty \frac{\|e_k\|}{\delta_k} \epsilon_k < \frac{a-1}{a+1} .$$

Then $\bar{g}$ is a minimal sequence which is a-equivalent to $\bar{e}$ whenever $\bar{g} = \{g_k\}_{k=1}^\infty$ is a sequence of elements in E with $\|g_k - e_k\| \le \epsilon_k$ for all k. If $\bar{e}$ is a basic sequence in E then $\bar{g}$ is a basic sequence, too.
(b) For any $\alpha > 0$ there is $\bar{\epsilon} = \{\epsilon_k\}_{k=1}^\infty$ such that for any sequence $\bar{g} = \{g_k\}_{k=1}^\infty$ in E with $\|e_k - g_k\| \le \epsilon_k$, $k = 1, 2, \ldots$, we have $\Theta(\text{ span}(\bar{e}), \text{span}(\bar{g})) < \alpha$.

Proof. We prove (a). (b) can be shown with similar arguments.

Let $0 < \epsilon < 1$ be such that $(1+\epsilon)(1-\epsilon)^{-1} < a$. Assume that $\bar{e}$ is $\bar{\delta}$-minimal with $\bar{\delta} = \{\delta_k\}_{k=1}^\infty$. Fix $\epsilon_k > 0$ such that $\sum_{k=1}^\infty \|e_k\|\epsilon_k/\delta_k < \epsilon$ and take a sequence $\bar{g}$ with $\|e_k - g_k\| \le \epsilon_k$ for all k. Consider $x = \sum_{k=1}^n \alpha_k e_k$ and define $Tx = \sum_{k=1}^n \alpha_k g_k$. Then we obtain, using 2.2.2,

$$\|x - Tx\| \le \sum_{k=1}^n |\alpha_k|\epsilon_k \le \left(\sum_{k=1}^n \frac{\|e_k\|}{\delta_k} \epsilon_k \right) \|x\| \le \epsilon \|x\| .$$

Hence $(1 - \epsilon)\|x\| \le \|Tx\| \le (1 + \epsilon)\|x\|$. This implies that T can be extended to an isomorphism T : closed span$(\bar{e}) \rightarrow$ closed span$(\bar{g})$ with $\|T\| \cdot \|T^{-1}\| \le (1+\epsilon)/(1-\epsilon) \le a$. As a consequence of 2.2.2 the sequence $\bar{g}$ is minimal, too. If $\bar{e}$ is a basic sequence then $\bar{g}$ is a basic sequence which follows, for example, from 2.3.1.   □

Sequences $\bar{e}$ and $\bar{g}$ with $\|e_k - g_k\| \le \epsilon_k$ for all k will be called $\bar{\epsilon}$-*close.*

There is a similar result for the stability of complete sequences. We have

Theorem 2.7.3 *[55, 56] Let $\bar{e} = \{e_k\}_{k=1}^\infty$ be a normalized $\bar{c}$-complete sequence for some sequence $\bar{c} = \{c_k\}_{k=1}^\infty$ of positive numbers. If for some sequence $\bar{\epsilon} = \{\epsilon_k\}_{k=1}^\infty$ of positive numbers the condition $\sum_k c_k \epsilon_k < 1$ holds then each sequence $\bar{g}$ which is $\bar{\epsilon}$-close to $\bar{e}$ is complete in E, too.*

Proof. Let $\bar{g}$ be $\bar{\epsilon}$-close to $\bar{e}$. Fix $\epsilon > 0$ such that $\epsilon < 1 - \sum_k c_k \epsilon_k$. Assume that $X := \text{closed span}(\bar{g}) \ne E$. Then $E/X \ne \{0\}$. Hence we find $y \in E$ with

$$1 = \|y + X\| = \inf_{x \in X} \|y - x\| \quad \text{and} \quad \|y\| \le 1 + \epsilon .$$

Since $\bar{e}$ is $\bar{c}$-complete there are α_k, $k = 1, 2, \ldots, n$ for suitable n, such that

$$\left\| y - \sum_{k=1}^n \alpha_k e_k \right\| \le \frac{\epsilon^2}{2} \quad \text{and} \quad |\alpha_k| \le c_k \|y\| \quad \text{for all } k .$$

We obtain, with $x_0 = \sum_{k=1}^n \alpha_k g_k$,

$$||y - x_0|| \leq \frac{\epsilon^2}{2} + \sum_{k=1}^{n} |\alpha_k| \cdot ||g_k - e_k||$$

$$\leq \frac{\epsilon^2}{2} + \sum_{k=1}^{n} c_k \epsilon_k ||y||$$

$$\leq \frac{\epsilon^2}{2} + 1 - \epsilon^2 < 1$$

contradicting $\inf_{x \in X} ||y - x|| = 1$. So X must be equal to E. □

We conclude this section with the following theorem due to Paley and Wiener, [115].

Theorem 2.7.4 *Let $\bar{e} = \{e_k\}_{k=1}^{\infty}$ be an orthonormal basis in a Hilbert space H and assume that $0 \leq \theta < 1$. If the sequence $\bar{g} = \{g_k\}_{k=1}^{\infty} \subset H$ satisfies the condition*

$$\left|\left|\sum_k \alpha_k (g_k - e_k)\right|\right|^2 \leq \theta \sum_k |\alpha_k|^2$$

for all scalars α_k then $\bar{g}$ is a Riesz basis of H, i.e. a basis of H which is equivalent to an orthonormal basis.

Proof. Put $T(\sum_{k=1}^{\infty} \alpha_k e_k) = \sum_{k=1}^{\infty} \alpha_k g_k$. From our assumption we obtain that T is well-defined and that

$$||x - Tx|| \leq \theta ||x|| \qquad \text{for all } x \in H.$$

Hence $S := id + \sum_{k=1}^{\infty} (id - T)^k$ converges in the operator norm and S satisfies $TS = ST = id$. This means that $T : H \to H$ is an onto-isomorphism. Thus $\bar{g}$ is a basis of H which is $||T|| \cdot ||T^{-1}||$-equivalent to the orthonormal basis $\bar{e}$. □

3

Isomorphisms, Isometries and Embeddings

Problems of isomorphisms and isometries play a central role in the theory of normed spaces. While two isomorphic normed spaces are identical from the linear point of view two isometric normed spaces are identical from a geometric point if view.

In this chapter we discuss basic notions such as almost isometries and Banach-Mazur distance. We also mention the largely unresolved connection between Banach-Mazur distance and projection constant. Finally we deal with polyhedral spaces and limiting spaces.

3.1 Classical Isomorphisms

We start with basic definitions.

Definition 3.1.1 *Two normed spaces E_1 and E_2 are called a-isomorphic for some $a \geq 1$ if there is an onto-isomorphism $T : E_1 \rightarrow E_2$ with*

$$||T|| \cdot ||T^{-1}|| \leq a .$$

Then T is called an a-isomorphism. E_1 and E_2 are almost isometric if they are $(1 + \epsilon)$-isomorphic for all $\epsilon > 0$.

There are well-known examples of almost isometric Banach spaces which are not isometric. The following example was communicated to us by P. Enflo.

Example. Let p_j be positive numbers with $p_j \uparrow \infty$. Put

$$E = \left(\sum_{j=1}^{\infty} \oplus l_{p_j}^2 \right)_{(l_2)} \quad \text{and} \quad F = (l_\infty^2 \oplus E)_{(l_2)} .$$

Then E satisfies $||x + y|| < ||x|| + ||y||$ whenever $x, y \in E$ are linearly independent. This is not true for F in view of the first component of F. Hence E

and F are not isometric. On the other hand, fix $\epsilon > 0$. Then there is j_0 such that there exist isomorphisms $T_j : l^2_{p_{j+1}} \to l^2_{p_j}$, $j \geq j_0$ and $T_0 : l^2_{p_j} \to l^2_\infty$ with $||T_j|| \cdot ||T_j^{-1}|| \leq 1 + \epsilon$, $j = 0$ and $j \geq j_0$. Define $T : E \to F$ by

$$T\{x_j\}_{j=1}^\infty = (T_0 x_{j_0}, x_{j_1}, \ldots, x_{j_0-1}, T_{j_0} x_{j_0+1}, T_{j_0+1} x_{j_0+2}, \ldots) .$$

Then we obtain $||T|| \cdot ||T^{-1}|| \leq 1 + \epsilon$. This proves, since $\epsilon > 0$ was arbitrary, that E and F are almost isometric.

There are even examples, due to Benyamini, of Banach spaces X and Y such that X^* and Y^* are isometrically isomorphic to l_1 and X and Y are almost isometric but not isometric.

However for Hilbert spaces we have

Proposition 3.1.2 *Let E and F be almost isometric Banach spaces. If F is a Hilbert space then E is a Hilbert space.*

Proof. Let B_E be the closed unit ball of E. Fix a sequence of $(1 + 1/n)$-isomorphisms $T_n : E \to F$ and consider the elements

$$\gamma_n = \{\langle T_n x, T_n y \rangle\}_{x,y \in B_E}$$

in $K = [-2, 2]^{B_E \times B_E}$. Since K is compact we find an accumulation point $\gamma \in K$ of $\{\gamma_n\}_{n=1}^\infty$, say $\gamma = \{\gamma(x,y)\}_{x,y \in B_E}$. For arbitrary $x, y \in E$ put $\langle\langle x, y \rangle\rangle = 0$ if $x = 0$ or $y = 0$ and

$$\langle\langle x, y \rangle\rangle = ||x|| \cdot ||y|| \cdot \gamma \left(\frac{x}{||x||}, \frac{y}{||y||} \right)$$

if $x \neq 0$ and $y \neq 0$. It is easily checked that $\langle\langle \cdot, \cdot \rangle\rangle$ is a scalar product in E with $||x|| = \sqrt{\langle\langle x, x \rangle\rangle}$ for all $x \in E$. $\square$

There are other Banach spaces different from Hilbert spaces which have a corresponding property. For example, let G be the space of almost universal disposition which we shall discuss in the next chapter. Then any Banach space E which is almost isometric to G is already isometric to G. (This is a consequence of 4.3.3.)

It is a well-known fact that all separable infinite dimensional Hilbert spaces are isometric.

If T is an isomorphism (isometry) from E_1 onto a subspace X of E_2 we call T an isomorphic (isometric) *embedding* of E_1 into E_2. X is then called isomorphic (isometric) *copy* of E_1.

The following amazing fact is due to Milutin which we mention without proof.

Theorem 3.1.3 *[108] For any metric compact space K whose cardinality is that of the continuum the space $C(K)$ is isomorphic to $C[0,1]$.*

Here we prove a classical fact concerning $C(K)$-spaces.

Proposition 3.1.4 *Let K be a metric compact space, fix $k \in K$ and put*

$$C_k(K) = \{f \ : \ f \in C(K), \ f(k) = 0\} \ .$$

Then $C_k(K)$ is always isomorphic to $C(K)$ provided that $C(K)$ is infinite dimensional. In particular, $C_0[0,1]$ is isomorphic to $C[0,1]$.

Proof. If $\dim C(K) = \infty$ then K has infinitely many points. Let d be the metric on K. If k is the only non-isolated point in K then $C(K)$ is isomorphic to c and $C_k(K)$ is isomorphic to c_0 and hence both spaces are isomorphic. So assume that there is a non-isolated point $h \in K \setminus \{k\}$. Using induction we define a sequence $\{f_j\}_{j=1}^{\infty}$ of elements in $C_k(K)$ vanishing at h.

Consider a closed infinite subset H_1 of K with diameter less than one containing h and fix an element $k_1 \notin H_1 \cup \{k\}$. Find a norm one element $f_1 \in C_k(K)$ with $f_1|_{H_1} = 0$ and $f_1(k_1) = 1$.

Assume next that we have already $k_1, \ldots, k_{n-1} \in K$, a closed infinite subset $H_{n-1} \subset K$ with diameter less than $(n-1)^{-1}$ containing h and norm one elements $f_j \in C_k(K)$, $j = 1, \ldots, n-1$, with

$$f_j(k_l) = \begin{cases} 1, j = l \\ 0, j \neq l \end{cases}, \quad f_j|_{H_{n-1}} = 0, \ l, j = 1, \ldots, n-1 \ ,$$

and $\sum_{j=1}^{n-1} |f_j| \leq 1 + \sum_{j=1}^{n-1} 2^{-j}$. Then consider a closed infinite subset $H_n \subset H_{n-1}$ with diameter less than $1/n$ containing h and fix an element $k_n \in H_{n-1} \setminus (H_{n-1} \cup \{k\})$. Put $A = \{s : \ s \in K, \ \sum_{j=1}^{n-1} |f_j(s)| \geq 2^{-n}\}$. In particular we have $k_1, \ldots, k_{n-1} \in A$. Find, using the Tietze extension theorem, $f_n \in C_k(K)$ with $f_n(k_n) = 1 = ||f_n||$ and $f_n|_{A \cup H_n} = 0$. Then we obtain $\sum_{j=1}^{n} |f_j| \leq 1 + \sum_{j=1}^{n} 2^{-j}$ on K and

$$f_j(k_l) = \begin{cases} 1, l = j \\ 0, l \neq j \end{cases}, \quad l, j = 1, \ldots, n \ .$$

Hence

$$\sup_j |\alpha_j| \leq \left\| \sum_{j=1}^{n} \alpha_j f_j \right\| \leq 2 \sup_j |\alpha_j|$$

for all n and all choices of α_j. Moreover, since $\lim_{n \to \infty} \operatorname{diam} H_n = 0$ we have $\lim_{n \to \infty} k_n = h$.

Put $X = $ closed span $\{f_j\}_{j=1}^{\infty}$. Then X is isomorphic to c_0. Moreover, we have $X \subset C_k(K)$ and X is complemented in $C(K)$. In fact, for $f \in C(K)$ define

$$Pf = \sum_{j=1}^{\infty} (f(k_j) - f(h)) f_j \ .$$

Since $\lim_j k_j = h$, P is a well-defined projection onto X with $||P|| \leq 2$. Let $Y = (id - P) C_k(K)$. Since $C_k(K)$ is complemented in $C(K)$ and $\mathbf{R} \oplus c_0$ is isomorphic to c_0 we obtain

$$C(K) = \mathbf{R} \oplus C_k(K) = \mathbf{R} \oplus X \oplus Y \sim \mathbf{R} \oplus c_0 \oplus Y \sim c_0 \oplus Y \sim C_k(K) \, .$$

Here " $\sim$ " means "is isomorphic to." □

3.2 The Banach-Mazur Distance

For any two isomorphic Banach spaces E_1 and E_2 put

$$d(E_1, E_2) = \inf\{\|T\| \cdot \|T^{-1}\| \ : \ T : E_1 \to E_2 \text{ an onto-isomorphism}\} \, .$$

Definition 3.2.1 $d(E_1, E_2)$ *is called the* Banach-Mazur distance *of E_1 and E_2.*

Originally Banach and Mazur studied $\log d(E_1, E_2)$, [3], which has the properties of a pseudometric but it is more convenient to deal with $d(E_1, E_2)$ instead, [53]. (We have $\log d(E_1, E_2) = 0$ whenever E_1 and E_2 are almost isometric.)

Let us give (without proof) several exact or approximate values of Banach-Mazur distances for finite dimensional spaces.

Theorem 3.2.2 *[69] Let E be an n-dimensional Banach space. Then we have*

$$d(l_2^n, E) \le \sqrt{n}, \qquad d(l_\infty^n, E) \le n,$$

$$d(l_{p_1}^n, l_{p_2}^n) = n^{|1/p_1 - 1/p_2|} \quad \text{if} \ \ 1 \le p_1, \ p_2 \le 2 \ \text{or} \ 2 \le p_1, \ p_2 < \infty \, ,$$

$$d(l_{p_1}^n, l_{p_2}^n) \sim \max\{n^{|1/2 - 1/p_1|}, n^{|1/2 - 1/p_2|}\} \quad \text{if} \ \ (p_1 - 2)(p_2 - 2) < 0 \, .$$

With Proposition 1.5.6 we obtain

Proposition 3.2.3 *The Banach-Mazur distance satisfies the multiplicative triangle inequality*

$$d(E_1, E_3) \le d(E_1, E_2) \cdot d(E_2, E_3)$$

for any isomorphic Banach spaces E_1, E_2 and E_3. Furthermore we have the following relation between the Banach-Mazur distance and the projection constant

$$\lambda(E_1) \le \lambda(E_2) \cdot d(E_1, E_2) \, .$$

We finish this section with a fundamental conjecture concerning the relationship between the projection constant of a finite dimensional Banach space and its Banach-Mzur distance to an l_∞^n-space.

Conjecture. *There exists a real-valued continuous function $f(t)$ on $[1, \infty[$ with $f(1) = 1$ such that, for any n-dimensional Banach space X, we have*

$$d(X, l_\infty^n) \le f(\, \lambda(X)) \, .$$

Only a few partial results related to this question are known. In 1.5.3 we showed that the conjecture is true in $t = 1$. This can be extended to values of t close to 1 which we mention without proof:

Theorem 3.2.4 *(See [145]) For some constant $\lambda_0 > 1$ there exists a function on $[1, \lambda_0]$ satisfying the preceding conjecture on this interval.*

Some more results suggesting a positive answer to the preceding conjecture were obtained by Bourgain, [7].

3.3 Minkowski Representation of n-Dimensional Banach Spaces and Limiting Spaces

Let $\{u_k\}_{k=1}^n$ be the unit vector basis of $\mathbf{R}^n$, i.e. $u_k = \underbrace{(0, \ldots, 0}_{k-1}, 1, 0, \ldots, 0)$.

Furthermore let E be any n-dimensional Banach space with normalized basis $\bar{e} = \{e_k\}_{k=1}^n$. We introduce a new norm $\|\cdot\|_{(E,\bar{e})}$ in $\mathbf{R}^n$ as follows. For each vector $x = \sum_{k=1}^n \alpha_k u_k$ we put

$$\|x\|_{(E,\bar{e})} = \left\| \sum_{k=1}^n \alpha_k e_k \right\| .$$

$\mathbf{R}^n_{(E,\bar{e})} := (\mathbf{R}^n, \|\cdot\|_{(E,\bar{e})})$ is called the *Minkowski representation* of $(E, \bar{e})$. This way we have defined a canonical isometry $T : E \to \mathbf{R}^n_{(E,\bar{e})}$ with $T(e_k) = u_k$ for all k. The unit sphere in $\mathbf{R}^n_{(E,\bar{e})}$ is called the *Minkowski surface* of $(E, \bar{e})$ (and *Minkowski curve* if $n = 2$).

In the following we always identify E with a Minkowski representation. Then the norm of each n-dimensional Banach space corresponds to a norm on $\mathbf{R}^n$.

Proposition 3.3.1 *The set $\mathcal{M}^n$ of all n-dimensional Banach spaces is a compact space with respect to the pseudometric $\log d(\cdot, \cdot)$ where $d(\cdot, \cdot)$ is the Banach-Mazur distance.*

Proof. Take $E_m \in \mathcal{M}^n$, $m = 1, 2, \ldots$. According to our remarks preceding 3.3.1 we can E_m identify with $(\mathbf{R}^n, \|\cdot\|_m)$ where $\|\cdot\|_m$ is a certain norm on $\mathbf{R}^n$. We can even choose an Auerbach system $\bar{e} = \{e_k\}_{k=1}^n$ in E_m (see 2.2.9) and arrange $\|\cdot\|_m$ such that

$$\|(\alpha_1, \ldots, \alpha_n)\|_m = \left\| \sum_{k=1}^n \alpha_k e_k \right\| \quad \text{for all} \quad \alpha_k .$$

(Recall that $\log d(E_m, \tilde{E}_m) = 0$ if $\tilde{E}_m$ is isometric to E_m). Let

$$B = \{(\alpha_1, \ldots, \alpha_n) \in \mathbf{R}^n \ : \ \max_k |\alpha_k| \le 1\} .$$

Then B is compact and

$$B_m := \{x \in \mathbf{R}^n \ : \ ||x||_m \leq 1\} \subset B \, .$$

Consider the element $\gamma_m = \{||x||_m\}_{x \in B}$ in $[0,1]^B$. Since $[0,1]^B$ is compact we find an accumulation point $\gamma = \{\gamma(x)\}_{x \in B}$ of $\{\gamma_m\}_{m=1}^{\infty}$. We easily check that

$$|||x||| = \begin{cases} \lambda\gamma(\frac{x}{\lambda}) & \text{if } x \in \lambda B \text{ and } x \neq 0 \\ 0 & \text{if } x = 0 \end{cases}$$

defines a seminorm on $\mathbf{R}^n$ with

$$\max_k |\alpha_k| \leq |||(\alpha_1, \ldots, \alpha_n)||| \quad \text{for all } \alpha_k \, .$$

Hence $||| \cdot |||$ is a norm on $\mathbf{R}^n$. Put $E = (\mathbf{R}^n, ||| \cdot |||)$. We obtain that, for each $\epsilon > 0$ and $m_0 > 0$, there is $m \geq m_0$ such that $\sup_{|||x||| \leq 1} |\ |||x||| - ||x||_m\ | \leq \epsilon$. Hence $\log d(E, E_m) \leq \epsilon$ which implies that $\mathcal{M}^n$ is compact. $\qquad\square$

We call $\mathcal{M}^n$ the *Minkowski compact*. Since compact metric spaces are separable we obtain from 3.3.1 a sequence of n-dimensional Banach spaces E_m such that, for any n-dimensional Banach space E and any $\epsilon > 0$, there is a suitable index m_0 with

$$d(E, E_{m_0}) \leq 1 + \epsilon$$

Definition 3.3.2 *A (not necessarily finite dimensional) Banach space X is called* polyhedral *if the Minkowski curves of any two-dimensional subspace are polygons.*

Clearly, if Y is a two-dimensional Banach space and a Minkowski curve of Y (with respect to a given normalized basis in Y) is a polygon then the Minkowski curve of Y with respect to any other normalized basis of Y is a polygon.

Proposition 3.3.3 *A Banach space X is polyhedral if and only if, for each linearly independent x and y in X, there exist $\alpha_x > 0$ and $\beta_y > 0$ such that $||\alpha x + \beta y||$ is additive in α and β on each of the four domains $(\alpha, \beta) \in]0, \alpha_x[\times]0, \beta_y[, (\alpha, \beta) \in]-\alpha_x, 0[\times]0, \beta_y[, (\alpha, \beta) \in]0, \alpha_x[\times]-\beta_y, 0[$ and $(\alpha, \beta) \in]-\alpha_x, 0[\times]-\beta_y, 0[$.*

Proof. The condition indicated is certainly equivalent to the additivity of $||\alpha x + \beta y||$ on the domains of the form $]\alpha_0, \alpha_0 + \theta_1\alpha_1[\times]\beta_0, \beta_0 + \theta_2\beta_1[$, for any $\theta_1, \theta_2 \in \{1, -1\}$ and any $\alpha_0, \beta_0 \in \mathbf{R}$ with respect to suitable α_1 and β_1 (in view of the homogeneity of the norm, consider $\alpha_0 x$ and $\beta_0 y$ instead of x and y). Hence the condition in Proposition 3.3.3 is equivalent to the situation where

$$\left\{ (\alpha, \beta) \in \mathbf{R}^2 \ : \ \left\| \alpha\frac{x}{||x||} + \beta\frac{y}{||y||} \right\| = 1 \right\}$$

is a polygon. Then the proposition is a direct consequence of the preceding definition. $\qquad\square$

Let Ω be a metric space. We consider a family $\mathcal{F}_\Omega = \{(E_\omega, \bar{e}_\omega)\}_{\omega \in \Omega}$ where E_ω are n-dimensional Banach spaces and $\bar{e}_\omega$ are normalized bases of E_ω. $\mathcal{F}_\Omega$ is called *totally minimal* with respect to ρ, for some $\rho > 0$, if each $\bar{e}_\omega$ is ρ-minimal (see 1.2.8). We denote the norm of the Minkowski representation of $(E_\omega, \bar{e}_\omega)$ in $\mathbf{R}^n$ by $\|\cdot\|_\omega$.

$\mathcal{F}_\Omega$ is called *converging* in $\omega \in \Omega$ if, for each sequence $\bar{\omega} = \{\omega_k\}_{k=1}^\infty \subset \Omega$ with $\lim_{k \to \infty} \omega_k = \omega$ and for each $x \in \mathbf{R}^n$, $\lim_{k \to \infty} \|x\|_{\omega_k}$ exists and does not depend on $\bar{\omega}$. We denote this limit by $\|x\|_{(\omega)}$. Then, obviously, $\|\cdot\|_{(\omega)}$ is a seminorm on $\mathbf{R}^n$. Let us point out that $\|\cdot\|_{(\omega)}$ is not necessarily equal to $\|\cdot\|_\omega$. $\|\cdot\|_{(\omega)}$ will be called *limiting (semi-)norm*.

Example. Let $\Omega = [0,1]$, $\bar{e}_0 = \{1, t\} \subset C[0,1]$ and $\bar{e}_\omega = \{\cos(\omega t), \sin(\omega t)\} \subset C[0,1]$ if $0 < \omega \leq 1$. Here we have

$$\|(\alpha, \beta)\|_\omega = \sup_{t \in [0,1]} |\alpha \cos(\omega t) + \beta \sin(\omega t)| \quad \text{if } 0 < \omega \leq 1$$

and $\|(\alpha, \beta)\|_0 = \sup_{t \in [0,1]} |\alpha + \beta t|$. Hence $\|\cdot\|_{(0)}$ exists but it is different from $\|\cdot\|_0$. $\|\cdot\|_{(0)}$ is not a norm since we obtain $\lim_{\omega \to 0} \|(0,1)\|_\omega = 0$.

Theorem 3.3.4 *Let the family $\mathcal{F}_\Omega$ be converging at $\omega \in \Omega$. Then the limiting seminorm $\|\cdot\|_{(\omega)}$ is a norm provided that $\mathcal{F}_\Omega$ is totally minimal.*

Proof. Let $\mathcal{F}_\Omega$ be totally minimal with respect to some $\rho > 0$. For each $x = (\alpha_1, \ldots, \alpha_n) \neq 0$ there is at least one k_0 with $\alpha_{k_0} \neq 0$. If $\omega_k \in \Omega$ and $\lim_{k \to \infty} \omega_k = \omega$ then we have, by 2.2.2,

$$\|x\|_{\omega_j} = \|(\alpha_1, \ldots, \alpha_n)\|_{\omega_j} \geq \rho |\alpha_{k_0}| > 0, \quad j = 1, 2, \ldots .$$

We obtain for the limiting seminorm

$$\|x\|_{(\omega)} \geq \rho |\alpha_{k_0}| > 0 .$$

Therefore $\|\cdot\|_{(\omega)}$ is a norm. □

The converse of 3.3.4 is also true in the following sense:

If $\|\cdot\|_{(\omega)}$ exists and is a norm then there is a neighbourhood U of ω in Ω such that $\{(E_\tau, \bar{e}_\tau)\}_{\tau \in U}$ is totally minimal.

This is again an easy consequence of the definitions.

$\mathbf{R}^n$ endowed with a limiting norm $\|\cdot\|_{(\omega)}$ will be called *limiting space* of the family $\mathcal{F}_\Omega$.

If $\|\cdot\|_{(\omega)}$ exists and is a norm then we obtain for the Banach-Mazur distances, if $\lim_{k \to \infty} \omega_k = \omega$,

$$\lim_{k \to \infty} \log d((\mathbf{R}^n, \|\cdot\|_{\omega_k}), (\mathbf{R}^n, \|\cdot\|_{(\omega)})) = 0 .$$

The notion of limiting spaces will be used in Chap. 12, for studying limiting Müntz spaces.

4

Spaces of Universal Disposition

A Banach space U is called *universal* (for all separable Banach spaces) if for each separable Banach space X there is a subspace Y in U such that X is isometric to Y. Banach and Mazur showed [3] that $C = C[0,1]$ is a universal space. Pelczynski proved the same result for the disk algebra A = closed span $\{z^k : k = 0, 1, \ldots\} \subset C_{\mathbf{C}}(\mathbf{T})$ where $\mathbf{T} = \{z \in \mathbf{C} : |z| = 1\}$ and $C_{\mathbf{C}}(\mathbf{T})$ is the space of all complexvalued continuous functions on $\mathbf{T}$.

Here we want to study spaces U which are universal with respect to embeddings of Banach spaces $X \subset Y$ where X and Y belong to special classes.

4.1 Coincidence of Embeddings

Let U be a Banach space and E a subspace of U. Furthermore let $\mathcal{X}$ be a given class of Banach spaces, $\mathcal{F}$ the class of all finite dimensional spaces and $\mathcal{S}$ the class of all separable Banach spaces.

Definition 4.1.1 *We shall call U a-universal relative to $\mathcal{X}$ if for any $X \in \mathcal{X}$ we can find a subspace F in U and an isomorphism T from X onto F with* $\max\{\|T\|, \|T^{-1}\|\} \leq a$. *If U is a-universal relative to $\mathcal{X}$ for all $a > 1$ (resp., for $a = 1$) we call U almost universal (resp.,* universal) *relative to $\mathcal{X}$.*

For example, $C = C[0,1]$ is universal relative to $\mathcal{S}$ and A is universal for the class of all separable Banach spaces over $\mathbf{C}$. However, here we only deal with the real case.

Proposition 4.1.2 c_0 *is almost universal relative to $\mathcal{F}$.*

Proof. Fix $\epsilon > 0$. Let $X \in \mathcal{F}$. Since $\dim X < \infty$ the closed unit ball of X is compact and we find finitely many $x_1^*, \ldots, x_m^* \in X^*$ with $\|x_k^*\| \leq 1$ for all k such that

$$(1 - \epsilon)\|x\| \leq \sup_{k \leq m} |x_k^*(x)| \leq \|x\| \quad \text{for all } x \in X$$

Define $T : X \to c_0$ by $Tx = (x_1^*(x), \ldots, x_m^*(x), 0 \ldots)$. Then we have $||T|| \leq 1$ and $||T^{-1}|| \leq (1 - \epsilon)^{-1}$. $\qquad\qquad\qquad\qquad\qquad\qquad\qquad\qquad\qquad\qquad\qquad\quad$ □

Now we want to introduce a much stronger property.

Definition 4.1.3 *We shall call E a* subspace of a-universal disposition *in U relative to $\mathcal{X}$ if for any pair of Banach spaces $X \in \mathcal{X}$ and $Y \in \mathcal{X}$, where $X \subset Y$ and X is isomorphic to E, and for any isomorphism $T : X \to E$ there are a subspace $F \supset E$ of U and an onto-isomorphism $\tilde{T} : Y \to F$ such that*

$$\tilde{T}|X = T \quad and \quad ||\tilde{T}|| \leq a||T||, \quad ||\tilde{T}^{-1}|| \leq a||T^{-1}|| \, .$$

If E is a subspace of a-universal disposition in U relative to $\mathcal{X}$ for all $a > 1$ (resp., for $a = 1$) we shall call E a subspace of almost universal disposition *(resp., universal disposition) relative to $\mathcal{X}$.*

So we have the following diagram

$$
\begin{array}{ccc}
\mathcal{X} \ni X & \xrightarrow{\;\;T\;\;} & E \subset U \\
\cap & & \cap \\
\mathcal{X} \ni Y & \xrightarrow{\;\;\tilde{T}\;\;} & F \subset U
\end{array}
$$

In other words: E in U mimics almost exactly all possible embeddings defined by $\mathcal{X}$.

Definition 4.1.4 *We shall call U a* space of a-universal *(resp., of almost universal, universal) disposition relative to $\mathcal{X}$ if any subspace E in U is a subspace of a-universal (resp., almost universal, universal) disposition in U relative to $\mathcal{X}$.*

For example a Hilbert space is a space of universal disposition relative to the class of all Euclidean spaces. In the next sections we discuss spaces of almost universal disposition relative to $\mathcal{F}$.

4.2 Existence of Spaces of Almost Universal Disposition

The main theorems about existence and non-existence of universal spaces relative to $\mathcal{F}$ are the following.

Theorem 4.2.1 *[45] There exists a separable Banach space of almost universal disposition relative to $\mathcal{F}$.*

A proof for 4.2.1 will be given in the more general context of 4.3.2.

Recall that x is a *smooth point* of the Banach space X if there is exactly one element $x^* \in X^*$ with $||x^*|| = 1$ and $x^*(x) = ||x||$. It is well known that the set of all smooth points in a separable Banach space is a dense subset [103].

Theorem 4.2.2 *[45] There does not exist a separable Banach space of universal disposition relative to $\mathcal{F}$.*

Proof. Assume that U is a separable Banach space of universal disposition for $\mathcal{F}$. Then let $\Omega = \{x_n\}_{n=1}^{\infty}$ be dense in U. Assume that $E_1, F_1 \subset U$ are finite dimensional subspaces of U such that there is an isometry $T_1 : E_1 \to F_1$. Fix n and put $E_2 = \mathrm{span}\{x_1, \ldots, x_n\} + E_1$. Find an isometric extension $T_2 : E_2 \to U$ of T_1. Then put $F_2 = TE_2$ and $F_3 = \mathrm{span}\{x_1, \ldots, x_n\} + F_2$. Find an isometric extension $T_3^{-1} : F_3 \to U$ of T_2^{-1}. Continuing in this fashion (letting $n \to \infty$), by induction starting from $E_1 = F_1 = \{0\}$, we find a surjective isometric extension $T_\infty : U \to U$ of T_1. This implies that every $x \in U$ with $||x|| = 1$ is a smooth point of U. Indeed, let x_0 be a smooth point with $||x_0|| = 1$ and put $E_1 = \mathrm{span}\{x_0\}$. Take any other point $x \in U$ with $||x|| = 1$ and put $F_1 = \mathrm{span}\{x\}$. According to the preceding argument we find an onto-isometry $T : U \to U$ with $Tx_0 = x$. Hence x is a smooth point, too. On the other hand, let $B = \mathbf{R}^2$ be endowed with the norm $||(s,t)|| = \max(|s|, |t|)$. Put $A = \{(s,0) : s \in \mathbf{R}\}$. Fix $x \in U$ with $||x|| = 1$ and define $T(s,0) = sx$. Then $T : A \to U$ is an isometry which can be extended to an isometry $\tilde{T} : B \to U$ by assumption. Since $(1,1)$ is non-smooth in B, $\tilde{T}(1,1)$ must be non-smooth in U, a contradiction. $\qquad\square$

4.3 Uniqueness and Universality of Spaces of Almost Universal Disposition

We start with a technical

Lemma 4.3.1 *Let G be a Banach space such that, for any $E, F \in \mathcal{F}$ with $E \subset F$ and $\dim F/E = 1$, any isometry $T : E \to G$ and any $\epsilon > 0$ there is a linear extension $\tilde{T} : F \to G$ of T with $||T|| \cdot ||T^{-1}|| \leq 1 + \epsilon$. Then G is of almost universal disposition relative to $\mathcal{F}$.*

Proof. Let $X, Y \in \mathcal{F}$ such that $X \subset Y$ and assume that there is an isomorphism $S : X \to G$. We claim that, for any $\epsilon > 0$, we can extend S to an isomorphism $\tilde{S} : Y \to G$ with $||\tilde{S}|| \leq (1 + \epsilon)||S||$ and $||\tilde{S}^{-1}|| \leq (1 + \epsilon)||S^{-1}||$. Without loss of generality we can assume that $\dim Y/X = 1$.

Endow $Y \oplus G$ with the norm $||(y,g)|| = ||y|| + ||g||$ for $y \in Y$ and $g \in G$ and put $V = \{(x, -||S||^{-1}Sx) \in Y \oplus G : x \in X\}$. Define $i : G \to (Y \oplus G)/V$ by $i(g) = (0,g) + V$ and $j : Y \to (Y \oplus G)/V$ by $j(y) = (y,0) + V$. We obtain

$$||g|| \geq ||i(g)|| = \inf_{x \in X} \left(||x|| + \left\| g - \frac{S}{||S||}x \right\| \right) \geq ||g|| \text{ for all } g \in G .$$

Hence i is an isometry. Moreover, since $||S|| \cdot ||S^{-1}|| \geq 1$,

$$||y|| \geq ||j(y)|| = \inf_{x \in X} \left(||y - x|| + \frac{||Sx||}{||S||} \right) \geq \frac{||y||}{||S|| \cdot ||S^{-1}||} \text{ for all } y \in Y .$$

Finally, we have $j(x) = (x,0) + V = (0, ||S||^{-1}Sx) + V$ for all $x \in X$. Put $E = jX$ and $F = jY$. Then dim $F/E = 1$. Let $T : E \rightarrow G$ be defined by $Tjx = i^{-1}jx = ||S||^{-1}Sx$. Then T is an isometry and we find, for any $\epsilon > 0$, an isomorphic extension $\tilde{T} : F \rightarrow G$ with $\max(||\tilde{T}||, ||\tilde{T}^{-1}||) \leq 1 + \epsilon$. Define $\tilde{S}y = ||S||\tilde{T}jy$, for $y \in Y$. Then $\tilde{S}$ extends S and we obtain $||\tilde{S}|| \leq (1+\epsilon)||S||$. Finally we have

$$||\tilde{S}y|| \geq \frac{||S||}{||\tilde{T}^{-1}||}||j(y)|| \geq \frac{||y||}{||S^{-1}|| \cdot ||\tilde{T}^{-1}||}$$

which implies $||\tilde{S}^{-1}|| \leq ||S^{-1}|| \cdot ||\tilde{T}^{-1}|| \leq (1+\epsilon)||S^{-1}||$. $\square$

In the following theorem we consider a general L_1-space, i.e. the Banach space of (classes of) absolutely integrable functions $f : \Omega \rightarrow \mathbf{R}$, where (Ω, Σ, μ) is a given measure space, endowed with the norm $||f|| = \int_\Omega |f|d\mu$.

Theorem 4.3.2 *[141] Let X be a separable Banach space whose dual X^* is isometric to an L_1-space. Then there is a space $G \supset X$ of almost universal disposition relative to $\mathcal{F}$ and a contractive projection $P : G \rightarrow X$.*

Proof. [92] a) Let X be a separable L_1-predual space. It was shown in [80] that it satisfies the following intersection properties of balls:

Whenever $x_1, \ldots, x_n \in X$ and $r_1, \ldots, r_n > 0$ such that $||x_i - x_j|| \leq r_i + r_j$ for $i \neq j$ then there is an element $u \in X$ with $||u + x_i|| \leq r_i$ for all i. (Here $-u$ lies in the intersection of all closed balls with center x_i and radius r_i.)

For a general Banach space Y let $B(Y)$ be its closed unit ball. For $A \subset Y$ let absconv(A) be the absolutely convex hull of A.

b) Now fix an arbitrary separable Banach space Y and a countable dense subset Δ of Y. Let Ω be the collection of all finite sets $\{(y_k, \lambda_k)\}_{k=1}^n$ where $y_1, \ldots, y_n \in \Delta$, $n = 1, 2, \ldots$, and the λ_k are rational numbers such that

$$\left\| \frac{y_k}{\lambda_k} - \frac{y_j}{\lambda_j} \right\| \leq \frac{1}{\lambda_k} + \frac{1}{\lambda_j} \quad \text{whenever } k \neq j.$$

Then Ω is countable, say $\Omega = \{\omega_m\}_{m=1}^\infty$ with $\omega_m = \{(y_{m,k}, \lambda_{m,k})\}_{k=1}^{n_m}$. Let U be the space of all finite sequences $(\alpha_1, \ldots, \alpha_n, 0, \ldots)$ with $\alpha_k \in \mathbf{R}$, $n = 1, 2, \ldots$, and put $u_k = (\underbrace{0, \ldots, 0}_{k-1}, 1, 0, \ldots)$. Let

$$B_1 = \{(y, 0) \in Y \oplus U \ : \ ||y|| \leq 1\}$$

and define $\Gamma(Y)$ to be the space $Y \oplus U$ endowed with the gauge functional of

$$B_2 := \text{absconv}\left(B_1 \cup \bigcup_{m=1}^\infty \{(y_{m,k}, \lambda_{m,k}u_m) \ : \ k = 1, \ldots, n_m\} \right)$$

as norm. Identify $y \in Y$ with $(y, 0) \in \Gamma(Y)$ and $u \in U$ with $(0, u) \in \Gamma(Y)$. Hence Y becomes a subspace of $\Gamma(Y)$. We claim that (after the identification)

$B_2 \cap Y = B_1$ which shows that the identification of Y with a subspace in $\Gamma(Y)$ is an isometric embedding. To prove the claim it suffices to show that $\|\sum_{k=1}^{n_m} \alpha_k y_{m,k}\| \leq 1$ whenever $\sum_k |\alpha_k| \leq 1$ and $\sum_{k=1}^{n_m} \alpha_k \lambda_{m,k} = 0$. To this end embed Y in a suitable $C(K)$-space (for example with $K = B(Y^*)$) which is an L_1-predual and find $f \in C(K)$ with $\|f + \lambda_{m,k}^{-1} y_k\| \leq \lambda_{m,k}^{-1}$ for all k (which is possible since $\|\lambda_{m,k}^{-1} y_k - \lambda_{m,j}^{-1} y_j\| \leq \lambda_{m,k}^{-1} + \lambda_{m,j}^{-1}$ if $k \neq j$). Hence

$$\left\| \sum_{k=1}^{n_m} \alpha_k y_{m,k} \right\| = \left| \sum_{k=1}^{n_m} \alpha_k \lambda_{m,k} \left(\frac{y_{m,k}}{\lambda_{m,k}} + f \right) \right| \leq \sum_{k=1}^{n_m} |\alpha_k| \leq 1 \,.$$

Clearly, for every m,

$$B_2 \cap (Y + \operatorname{span}\{u_m\}) = \operatorname{absconv}(B_1 \cup \{(y_{m,k}, \lambda_{m,k} u_m) \,:\, k = 1, \ldots, n_m\}).$$

From now on, for $y \in Y$ and $u \in U$ we write $y + u \in \Gamma(Y)$ instead of $(y, u) \in \Gamma(Y)$.

If $P : Y \to X$ is a contractive linear map then we find $x_m \in X$ with $\|P(\lambda_{m,k}^{-1} y_{m,k}) + x_m\| \leq \lambda_{m,k}^{-1}$ for all k and m since X is an L_1-predual. Extend P to $\tilde{P} : \Gamma(Y) \to X$ by putting $\tilde{P}(u_m) = x_m$ for all m. Then $\tilde{P}$ becomes a contractive extension of P.

c) Now start with $Y_0 = X$ and put $Y_1 = \Gamma(Y_0)$. Repeat the procedure, put $Y_2 = \Gamma(Y_1)$, $Y_3 = \Gamma(Y_2)$ etc. and obtain, by induction, a sequence of spaces $Y_0 \subset Y_1 \subset \ldots$. Take G to be the completion of $\cup_{j=1}^{\infty} Y_j$. Clearly, by b), $P_0 = id$ on Y_0 can be extended to a contractive projection $P : G \to X$.

Finally we show that G is of almost universal disposition relative to $\mathcal{F}$. To this end let E, F be finite dimensional Banach spaces with $E \subset F$ and dim $F/E = 1$ and let $T : E \to G$ be an isometry. Fix $\epsilon > 0$ such that $\epsilon < 1/4$. Let $f_0 \in F \setminus E$ be such that there is a projection $Q : F \to \operatorname{span}\{f_0\}$ with $QE = \{0\}$ and $\|Q\| = 1$. (Find $f^* \in F^*$ with $\|f^*\| = 1$ and $f^*|_E = 0$. Then take $f_0 \in F$ with $f^*(f_0) = 1$.) Let $e_1, \ldots, e_n \in E$ and $\lambda_1, \ldots, \lambda_n \in \mathbf{Q}$ be such that $B_0 := \operatorname{absconv}(B(E) \cup \{\lambda_k f_0 + e_k \,:\, k = 1, \ldots, n\})$ satisfies

$$\frac{1}{1+\epsilon} B(F) \subset B_0 \subset B(F) \,.$$

Let $T_1 : E \to G$ be such that $T_1 E \subset Y_j$ for some j and $\|T_1\| \leq 1 + \epsilon$, $\|T - T_1\| \leq \epsilon$, $\|T_1^{-1}\| \leq 1 + \epsilon$. Hence

$$\frac{1}{1+\epsilon} B(T_1 E) \subset T_1 B(E) \subset (1 + \epsilon) B(T_1 E) \,.$$

In view of the construction in b) we can arrange the $T_1 e_k$ such that, in addition, for some $u \in Y_{j+1}$,

$$B(Y_{j+1} \cap (T_1 E + \operatorname{span}\{u\}))$$
$$= \operatorname{absconv}(B(T_1 E) \cup \{\lambda_k u + T_1 e_k \,:\, k = 1, \ldots, n\}).$$

Extend T_1 to $\tilde{T}_1 : F \to G$ by putting $\tilde{T}_1 f_0 = u$. This yields

$$\frac{1}{1+\epsilon} \tilde{T}_1 B(F) \subset \tilde{T}_1 B_0 \subset (1+\epsilon) B(\tilde{T}_1 F) \subset (1+\epsilon)^2 \tilde{T}_1 B(F)$$

and we have $||\tilde{T}_1|| \leq (1+\epsilon)^2$ and $||\tilde{T}_1^{-1}|| \leq (1+\epsilon)^2$. Finally, let

$$\tilde{T} = T(id - Q) + \tilde{T}_1 Q .$$

Then $\tilde{T}$ is an extension of T with $||\tilde{T} - \tilde{T}_1|| \leq 2\epsilon$. Hence $||\tilde{T}|| \leq (1+\epsilon)^2 + 2\epsilon$. Moreover, we obtain, for $f \in F$,

$$\frac{||f||}{(1+\epsilon)^2} \leq \frac{||f||}{||\tilde{T}_1^{-1}||} \leq ||\tilde{T}f|| + ||(\tilde{T} - \tilde{T}_1)f|| \leq ||\tilde{T}f|| + 2\epsilon ||f||$$

which implies

$$||\tilde{T}^{-1}|| \leq \left(\frac{1}{(1+\epsilon)^2} - 2\epsilon \right)^{-1} .$$

Thus we have

$$||\tilde{T}|| \cdot ||\tilde{T}^{-1}|| \leq ((1+\epsilon)^2 + 2\epsilon) \left(\frac{1}{(1+\epsilon)^2} - 2\epsilon \right)^{-1} .$$

Since $\epsilon > 0$ can be chosen arbitrarily small Lemma 4.3.1 completes the proof.
□

Let G_1 and G_2 be two separable spaces of almost universal disposition relative to $\mathcal{F}$. Using this property and induction we find, for every $\epsilon > 0$, a surjective $(1+\epsilon)$-isomorphism $T : G_1 \to G_2$. (This is the $(1+\epsilon)$-analogue of the proof of 4.2.2.) By refining this argument one can indeed show

Theorem 4.3.3 *[86] Two separable spaces of almost universal disposition relative to $\mathcal{F}$ are isometric.*

So we have essentially one separable Banach space of almost universal disposition relative to $\mathcal{F}$. This space will be denoted by G. It can be shown that G^* is isometric to an L_1-space, i.e. G is an L_1-predual space. Moreover G satisfies 4.3.2 and hence contains all separable L_1-predual spaces as complemented subspaces. In particular, G contains an isometric copy of C and is thus universal relative to $\mathcal{S}$.

Mazur formulated the following transitivity problem:

Does there exist a separable infinite dimensional non-Hilbert space X such that, for any x and y in X with $||x|| = ||y||$, there is an onto-isometry $T : X \to X$ with $Tx = y$?

The space G comes close to this property, it is "almost transitive " in the following sense.

Theorem 4.3.4 *[86, 90] For each x and y in G with $||x|| = ||y||$ and each $\epsilon > 0$ there is a $(1+\epsilon)$-isomorphism T from G onto G with $Tx = y$. If x and y both are smooth points of G then T can be chosen to be an isometry.*

Using the fact that G has almost universal disposition relative to $\mathcal{F}$ we find as before, for every $\epsilon > 0$ and all $x, y \in G$ with $||x|| = ||y|| = 1$ an $(1 + \epsilon)$-isomorphism T from G onto G with $Tx = y$. Again a (highly non-trivial) refinement of this argument proves the second part of 4.3.4.

Problems 4.3.5 *Does there exist a subspace X in c_0 of a-universal disposition for some a?*

4.4 Synthesis of Sequences

What we did with Banach spaces and dispositions of their subspaces in the preceding sections can also be done with sequences. There are notions of universality for bases. For example the follwing theorem is due to Pelczynski.

Theorem 4.4.1 *[117] (a) There is a Banach space U_1 with a universal basis $\bar{u}$ in the sense that, for each Banach space X with basis $\bar{e}$, there is a subsequence of $\bar{u}$ which is equivalent to $\bar{e}$.*
(b) There is a Banach space U_2 with a universal unconditional basis $\bar{v}$ in the sense that, for any Banach space X with unconditional basis $\bar{f}$, there is a subsequence of $\bar{v}$ which is equivalent to $\bar{f}$.

Proof. We use that every separable Banach space X can be isometrically embedded into $C[0, 1]$. Let $\{f_n\}_{n=1}^{\infty}$ be dense in $C[0, 1]$. For a sequence of real numbers α_k put

$$||\{\alpha_k\}_{k=1}^{\infty}||_1 = \sup_n \left\| \sum_{k=1}^{n} \alpha_k f_k \right\|$$

and

$$||\{\alpha_k\}_{k=1}^{\infty}||_2 = \sup\{|\textstyle\sum_{k=1}^{n} \alpha_{\pi(k)} f_{\pi(k)}:$$

$$\pi \text{ a permutation of the indices }, n = 1, 2, \ldots\}.$$

Define $U_1 = \{\{\alpha_k\}_{k=1}^{\infty} : ||\{\alpha_k\}_{k=1}^{\infty}||_1 < \infty\}$ and

$$U_2 = \{\{\alpha_k\}_{k=1}^{\infty} : ||\{\alpha_k\}_{k=1}^{\infty}||_2 < \infty\}.$$

Let $u_k = v_k = (\underbrace{0, \ldots, 0}_{k-1}, 1, 0, \ldots)$. Then $\bar{u} = \{u_k\}_{k=1}^{\infty}$ is a monotone basis for U_1 (see 2.3.1) and $\bar{v} = \{v_k\}_{k=1}^{\infty}$ is an unconditional basis for U_2. If $\bar{x} = \{x_k\}_{k=1}^{\infty}$ is any basis in any Banach space X we may regard $\bar{x}$ as sequence in $C[0, 1]$. Fix $\epsilon > 0$ and $\epsilon_k > 0$. Find f_{n_k} such that $||f_{n_k} - x_k|| \le \epsilon_k$ for all k. If the ϵ_k are small enough, according to 2.7.2, $\bar{x}$ and $\{f_{n_k}\}_{k=1}^{\infty}$ are $(1+\epsilon)$-equivalent in $C[0, 1]$. Since we have

$$\frac{1}{\nu(\bar{x})} \sup_n \left\| \sum_{k=1}^{n} \alpha_k x_k \right\| \le \left\| \sum_{k=1}^{\infty} \alpha_k x_k \right\| \le \sup_n \left\| \sum_{k=1}^{n} \alpha_k x_k \right\|$$

for all α_k, where $\nu(\bar{x})$ is the basis constant of $\bar{x}$, we see that $\bar{x}$ and $\{u_{n_k}\}_{k=1}^{\infty}$ are $\nu(\bar{x})(1 + \epsilon)$-equivalent with respect to the norm $\|\cdot\|_1$. Similarly, if $\bar{x}$ is unconditional we see that $\bar{x}$ and $\{v_{n_k}\}_{k=1}^{\infty}$ are equivalent with respect to $\|\cdot\|_2$.   □

The construction in the proof even includes that, for every $\epsilon > 0$ and every monotone basis, there is a suitable subsequence of $\bar{u}$ which is $(1+\epsilon)$-equivalent to $\bar{u}$. The preceding proof can be extended to show that for every basis $\bar{x}$ in X there is a subsequence $\{u_{n_k}\}_{k=1}^{\infty}$ of $\bar{u}$ which is equivalent to $\bar{x}$ such that the natural projection $P : U \to \overline{\text{span}}\{u_{n_k}\}_{k=1}^{\infty}$ with $P(\sum_j \alpha_j u_j) = \sum_k \alpha_{n_k} u_{n_k}$ for all α_j is bounded.

On the other hand there is a number of negative results about the existence of universal sequences. We conclude this section by giving one of them.

Theorem 4.4.2 *[38] An infinite dimensional separable Hilbert space does not have a basis $\bar{u}$ which is universal for all normalized bases in Hilbert spaces, i.e. such that all normalized bases in Hilbert spaces are equivalent to suitable subsequences of $\bar{u}$.*

5

Bounded Approximation Properties

Throughout this section let X be a separable Banach space. We investigate certain approximation properties for X and deal with the question under which additional condition then X has a basis. In this context we also give a sufficient criterion for X to be a dual Banach space.

5.1 Basic Definitions

We want to study some bounded approximation properties of X which might be regarded as generalizations of the notion of basis.

Definition 5.1.1 *(i)* X *is said to have the* bounded approximation property *(BAP) if there exists a sequence of linear bounded operators* $R_n : X \to X$ *of finite rank such that* $\lim_{n \to \infty} R_n x = x$ *for all* $x \in X$. *If in addition* $\|R_n\| = 1$ *for all* n *then* X *has the* metric approximation property *(MAP)*.

(ii) X *has the* commuting bounded approximation property *(CBAP) if* X *has the BAP with respect to a sequence* $\{R_n\}_{n=1}^\infty$ *satisfying the condition of (i) and, in addition,*
$$R_n R_m = R_{\min(m,n)} \quad \text{if} \quad m \neq n \,.$$
$\{R_n\}_{n=1}^\infty$ *is then called* commuting approximating sequence *(c.a.s.) of* X.

(iii) X *has the* finite dimensional decomposition property *(FDD) if there is a c.a.s.* $\{R_n\}_{n=1}^\infty$ *of* X *satisfying in addition* $R_n^2 = R_n$ *for all* n.

Note that, by the uniform boundedness principle, we obtain $\sup_n \|R_n\| < \infty$ whenever $\lim_{n \to \infty} R_n x = x$ for all $x \in X$. FDD means that we have $X = \sum_{n=1}^\infty \oplus((R_n - R_{n-1})X)$, since in this case the R_n are projections. (We always put $R_0 = 0$.) Moreover we have dim $(R_n - R_{n-1})X < \infty$ (see 2.6.) In this context the existence of a basis can be characterized as follows:

Lemma 5.1.2 X *has a basis if and only if* X *has an FDD with respect to a c.a.s.* $\{R_n\}_{n=1}^{\infty}$ *such that* $\sup_n \dim (R_n - R_{n-1})X < \infty$.

Proof. The necessity of the condition for the existence of a basis is clear. For the sufficiency put $m_n = \dim (R_n - R_{n-1})X$ and find bases $\{e_{n,j}\}_{j=1}^{m_n}$ of $(R_n - R_{n-1})X$ with uniformly bounded basis constants (for example Auerbach systems). Then $\{e_{n,j} : j = 1, \ldots, m_n; \ n = 1, 2, \ldots\}$ (in the lexicocraphical order of the indices) is a basis of X. (See 2.6.2.) $\qquad\square$

As a direct consequence of the definition we obtain the following line of implications:

$$X \text{ has a basis} \Rightarrow X \text{ has FDD} \Rightarrow X \text{ has CBAP} \Rightarrow X \text{ has BAP} \ .$$

On the other hand it is known [123, 135] that X has CBAP $\not\Rightarrow X$ has FDD and X has FDD $\not\Rightarrow X$ has a basis.

The fundamental fact concerning the connection of BAP and basis goes back to Pelczynski, [118], see also [64]:

Theorem 5.1.3 X *has the BAP if and only if there is a Banach space* V *such that* $X \oplus V$ *has a basis.*

Proof. If $X \oplus V$ has a basis $\bar{e} = \{e_k\}_{k=1}^{\infty}$ then let R_m be the basis projections of $\bar{e}$ (i.e. $R_m(\sum_k \alpha_k e_k) = \sum_{k=1}^{m} \alpha_k e_k$ for all α_k). Moreover let $P : X \oplus V \to X$ be the projection with ker $P = V$. Then $PR_m|_X$ are finite rank operators on X with $\lim_{m\to\infty} PR_m x = x$ for all $x \in X$.

Conversely, let $R_m : X \to X$ be finite rank operators with $\lim_{m\to\infty} R_m x = x$ for all $x \in X$. Let $n_m = \dim (R_m - R_{m-1})X$ and let $\{x_{m,j} : j = 1, \ldots, n_m\}$ be an Auerbach system of $(R_m - R_{m-1})X$ with conjugate system $\{x_{m,j}^*\}_{j=1}^{n_m}$. Let Z be the Banach space of all sequences $\{y_l\}_{l=1}^{\infty}$ with $y_j \in X$ and $\|\{y_l\}_{l=1}^{\infty}\| = \sup_n \|\sum_{l=1}^{n} y_l\| < \infty$. Fix an integer l, say $l = \sum_{m'=1}^{m-1} n_{m'}^2 + kn_m + j$ for some m, $k < n_m$ and $j < n_m$. Define $e_l = \{u_{l'}\}_{l'=1}^{\infty}$ with $u_{l'} = x_{m,j}$ if $l' = l$ and $u_{l'} = 0$ else. (The index l goes n_1-times through the indices of the Auerbach system of $(R_1 - R_0)X$, then n_2-times through that of $(R_2 - R_1)X$ and so on.) By definition of the norm in Z, since each e_l lives on a different component, $\bar{e} = \{e_l\}_{l=1}^{\infty}$ is a basic sequence in Z.

Let $S : Z \to X$ be the operator with $S\{y_l\}_{l=1}^{\infty} = \sum_{l=1}^{\infty} y_l$. Hence $\|S\| = 1$. Finally, define $T : X \to Z$ by

$$Tx = \{v_l\}_{l=1}^{\infty} \text{ with } v_l = \frac{1}{n_m} x_{m,j}^* ((R_m - R_{m-1})x) \, x_{m,j}$$

if $l = \sum_{m'=1}^{m-1} n_{m'}^2 + kn_m + j$. Then, using $\sum_{m'=1}^{m-1} (R_{m'} - R_{m'-1}) = R_{m-1}$ (with $R_0 = 0$), we obtain

$$\|Tx\| =$$

$$\sup_{m,k,j} \|R_{m-1}x + \frac{k}{n_m}(R_m - R_{m-1})x + \frac{1}{n_m}\sum_{j'=1}^{j} x^*_{m,j}\left((R_m - R_{m-1})x\right)x_{m,j}\|$$

$$\leq 5(\sup_n \|R_n\|)\|x\| .$$

Moreover $STx = x$ for all $x \in X$. Put $Y = $ closed span $(\bar{e})$. Then $\bar{e}$ is a basis of Y. TS is a bounded projection from Y onto TX. Hence we have $Y = TX \oplus (\ker S \cap Y)$ and TX is isomorphic to X. $\qquad \square$

Sometimes it is easier to show the existence of a basis in X by using indirect methods, see [100]:

Theorem 5.1.4 *Let X have the CBAP and assume there is a c.a.s. $\{R_n\}_{n=1}^{\infty}$ such that the $R_n - R_{n-1}$ factor uniformly through $l_p^{m_n}$'s for some p with $1 \leq p \leq \infty$. I.e. assume that there are uniformly bounded linear operators*

$$T_n : X \to l_p^{m_n} \quad and \quad S_n : l_p^{m_n} \to X$$

satisfying $S_n T_n = R_n - R_{n-1}$ for all n. Then X has a basis.

Proof. We prove Theorem 5.1.4 only under the additional assumption that X contains a complemented subspace Y which is isomorphic to l_p if $1 \leq p < \infty$ in the assertion of the theorem or to c_0 if $p = \infty$. Since Y is then isomorphic to $(Y \oplus Y \oplus \ldots)_{(l_p)}$ or to $(Y \oplus Y \oplus \ldots)_{(c_0)}$ we obtain that $X = Z \oplus Y$, for some subspace Z, is isomorphic to $Z \oplus Y \oplus Y = X \oplus Y$. So we have to show that $X \oplus Y$ has a basis. However, the theorem is true in general without this assumption (see [100]).

We show the case $1 \leq p < \infty$, the case $p = \infty$ is identical (where we have to replace l_p by c_0). Take S_n, T_n, and $D_n := l_p^{m_n}$ as in the statement of the theorem. Assume

$$S_n D_n \subset (R_{n+1} - R_{n-2})X \quad \text{for all } n \tag{5.1}$$

This is no loss of generality since otherwise take $(R_{n+1} - R_{n-2})S_n$ instead of S_n and use $(R_{n+1} - R_{n-2})(R_n - R_{n-1}) = R_n - R_{n-1}$. (Put $R_0 = R_{-1} = 0$.) Let

$$U = \text{closed span}\{\{(R_{k+1} - R_{k-2})x\}_{k=1}^{\infty} : x \in X, \text{ there is } n \text{ with } R_n x = x\}$$

be regarded as subspace of $(\sum_n \oplus R_{n+1}X)_{(l_p)}$. (Recall, if $R_n x = x$ then $(R_{k+1} - R_{k-2})x = 0$ for $k \geq n+3$). Put $V := (\sum \oplus D_k)_{(l_p)}$. Then we have $V = l_p$.

a) We claim that U is isomorphic to l_p. Indeed, let $S : V \to U$ be defined by

$$S\{d_j\}_{j=1}^{\infty} = \{(R_{k+1} - R_{k-2})\sum_j S_j d_j\}_{k=1}^{\infty} \quad \text{if } \{d_k\}_{k=1}^{\infty} \in V .$$

Then S is well-defined and bounded. (Consider d_k which are eventually zero and use (5.1). We have

$$(R_{k+1} - R_{k-2})(R_{j+1} - R_{j-2}) = 0$$

if $k \geq j + 4$ or $k \leq j - 4$.) Moreover, let $T : U \to V$ be the operator with

$$T\{(R_{k+1} - R_{k-2})x\}_{k=1}^{\infty} = \{T_k(R_{k+1} - R_{k-2})x\}_{k=1}^{\infty}, \quad x \in X$$

which is clearly bounded.

If $\{d_k\}_{k=1}^{\infty} \in TU$ where $d_k = T_k(R_{k+1} - R_{k-2})x$ for some x then we obtain

$$(R_{n+1} - R_{n-2})\sum_k S_k d_k = (R_{n+1} - R_{n-2})\sum_k (R_k - R_{k-1})x$$

$$= (R_{n+1} - R_{n-2})x.$$

Hence $ST = id_U$ which implies that $TS : V \to TU$ is a bounded projection and U is isomorphic to TU. Since $V = l_p$ the space TU as a complemented subspace of V is isomorphic to l_p [116]. This proves the claim.

b) Construction of an FDD on $X \oplus l_p$. Put

$$i(x) = \{(R_{k+1} - R_{k-2})x\}_{k=1}^{\infty} \in U \quad \text{if } x \in X \text{ and there is } n \text{ with } R_n x = x .$$

i is not bounded on X but i is uniformly bounded on all $(R_{n+1} - R_{n-2})X$ since $R_m(R_{n+1} - R_{n-2}) = 0$ if $m < n - 2$ and $(id - R_m)(R_{n+1} - R_{n-2}) = 0$ if $m > n + 1$.

Finally, let $\bar{R}_n : U \to U$ be defined by

$$\bar{R}_n\{(R_{k+1} - R_{k-2})x\}_{k=1}^{\infty} = \{(R_{k+1} - R_{k-2})R_n x\}_{k=1}^{\infty} .$$

Then $\bar{R}_n \bar{R}_m = \bar{R}_{\min(n,m)}$, if $n \neq m$, and $\bar{R}_n \to id$ pointwise on U. Moreover, we have $i \circ R_n = \bar{R}_n \circ i$.

Find finite dimensional $E_n \subset U$ with

$$i(R_{n+1}X) \subset E_n \quad \text{and} \quad \sup_n d(E_n, l_p^{\dim E_n}) < \infty$$

($d(\cdot, \cdot)$ is the Banach-Mazur distance, see 3.2.)

Put $W = X \oplus (\sum \oplus E_n)_{(l_p)}$ and define on W

$$P_n(x, \{e_k\}_{k=1}^{\infty}) = (R_n x + i^{-1}(\bar{R}_n - \bar{R}_n^{\,2})e_n, (e_1, \ldots, e_{n-1},$$

$$i(R_{n+1} - R_{n-1})x + (id - \bar{R}_n)e_n, 0, \ldots)) .$$

It is easily checked that $P_{3n}P_{3m} = P_{3\min(n,m)}$ for all n and m including the case $n = m$.

If $e_n \in E_n$ with $e_n = \{(R_{k+1} - R_{k-2})x\}_{k=1}^{\infty}$ for some $x \in X$ then

$$f := (\bar{R}_n - \bar{R}_n^2)e_n =$$

$$(\underbrace{0, \ldots, 0}_{n-2}, R_n^2(id - R_n)x, R_n(id - R_n)x, R_n(id - R_n)x, R_n(id - R_n)^2x, 0, \ldots)$$

and $i^{-1}(f) = (R_n - R_n^2)x$. Hence, $\|i^{-1}(f)\| \leq \|f\|$. This implies that $\sup_n \|P_n\| < \infty$. Thus, $\{P_{3n}\}_{n=1}^{\infty}$ is a c.a.s. of an FDD of W.

c) Construction of a basis in $W \oplus l_p$. At first we claim that $P_n - P_{n-1}$ factors uniformly through l_p^m-spaces. To this end put

$$G_n = D_{n-2} \oplus \ldots D_{n+2} \oplus E_{n-1} \oplus E_n$$

Define, $\tilde{S}_n : G_n \to W$, with $x = S_{n-2}d_{n-2} + \ldots + S_{n+2}d_{n+2}$, by

$$\tilde{S}_n(d_{n-2}, \ldots, d_{n+2}, e_{n-1}, e_n) =$$

$$((R_n - R_{n-1})x + i^{-1}(\bar{R}_n - \bar{R}_n^2)e_n - i^{-1}(\bar{R}_{n-1} - \bar{R}_{n-1}^2)e_{n-1}, \underbrace{(0, \ldots, 0}_{n-2},$$

$$\bar{R}_{n-1}e_{n-1} - i(R_n - R_{n-2})x, (id - \bar{R}_n)e_n + i(R_{n+1} - R_{n-1})x, 0, \ldots)) \, .$$

Moreover, let $\tilde{T}_n : W \to G_n$ be the operator with

$$\tilde{T}_n(x, \{e_k\}_{k=1}^{\infty}) = (T_{n-2}x, \ldots, T_{n+2}x, e_{n-1}, e_n) \, .$$

Then $P_n - P_{n-1} = \tilde{S}_n \tilde{T}_n$. Hence also the $P_{3n} - P_{3n-3}$ factor uniformly through l_p^m-spaces. Since the $P_{3n} - P_{3n-3}$ are projections it is easily checked that then the spaces $(P_{3n} - P_{3n-3})W$ are uniformly complemented in l_p^m-spaces. Hence there are finite dimensional Banach spaces $F_n \supset (P_{3n} - P_{3n-3})W$ with $\sup_n d(F_n, l_p^{\dim F_n}) < \infty$ and there exist uniformly bounded projections $Q_n : F_n \to (P_{3n} - P_{3n-3})W$. This implies that $F := (\sum_n \oplus(id - Q_n)F_n)_{(l_p)}$ is complemented in $(\sum_n \oplus F_n)_{(l_p)}$. The latter space is isomorphic to l_p. Therefore F, as a complemented subspace of l_p, is isomorphic to l_p. Hence we obtain

$$Z = X \oplus l_p \sim X \oplus l_p \oplus l_p \sim W \oplus F \sim \left(\sum_n \oplus Q_n F_n \right) \oplus \left(\sum_n \oplus(id - Q_n)F_n \right)_{(l_p)}$$

where " $\sim$ " means "is isomorphic to". Thus, Z has an FDD whose summands are uniformly isomorphic to $Q_n F_n \oplus (id - Q_n)F_n = F_n$ that is, to $l_p^{\dim F_n}$. Take in each F_n the elements corresponding to the unit vector basis of $l_p^{m_n}$. They form, in the lexicographical order, a basis of Z (see 2.6.2). $\qquad\square$

The proof does not contain arguments which depend on the real numbers. It carries over literally to the complex case.

Note that the condition of the theorem is also necessary for the existence of a basis in X. This follows from the fact that if R_n are basis projections on X then $R_n - R_{n-1}$ always has rank one and therefore factors through any l_p's. (Here $R_n(\sum_{k=1}^{\infty} \alpha_k e_k) = \sum_{k=1}^{n} \alpha_k e_k$.)

If we replace $R_n - R_{n-1}$ by R_n in the assumption of the theorem then we obtain a Banach space X which has finite dimensional subspaces $E_1 \subset E_2 \subset \ldots$ with $X = \overline{\cup_n E_n}$ and $\sup_n d(E_n, l_p^{m_n}) < \infty$ for some m_n, [82]. Such a space X is called a $\mathcal{L}_p$-space.

Finally we recall that it is an open question whether there exists a separable Banach space with non-standard projection type (see 1.6). Such a space cannot have FDD which follows from the definition in 1.6. (The proof is identical to that of 1.6.2.)

5.2 The Shrinking CBAP

Now we discuss special CBAP's.

Definition 5.2.1 *Let* $\{R_n\}_{n=1}^{\infty}$ *be a c.a.s. of* X *such that* $\{R_n^*\}_{n=1}^{\infty}$ *is a c.a.s. for* X^**. Then* $\{R_n\}_{n=1}^{\infty}$ *is called a* shrinking c.a.s. *and* X *is said to have the* shrinking CBAP.

If X *is isomorphic to a dual Banach space* Y^* *and* $\{R_n\}_{n=1}^{\infty}$ *is a c.a.s. of* Y *such that* $\{R_n^*\}_{n=1}^{\infty}$ *is a c.a.s. of* X *then* $\{R_n^*\}_{n=1}^{\infty}$ *is called a* w^*-shrinking c.a.s. *and* X *has a* w^*-shrinking CBAP.

If X is a reflexive Banach space with CBAP then every c.a.s. $\{R_n\}_{n=1}^{\infty}$ is shrinking. This is an easy consequence of the separation theorem.

In the following we always put $R_0 = 0$.

Theorem 5.2.2 *a) Assume that* X *is a closed subspace of* l_p, $1 < p < \infty$. *Then for every c.a.s* $\{R_n\}_{n=1}^{\infty}$ *there are integers* $0 = m_0 < m_1 < m_2 < \ldots$ *and numbers* $c_1 > 0$, $c_2 > 0$ *satisfying*

$$c_1 \left(\sum_k \|(R_{m_{k+1}} - R_{m_k})x\|^p \right)^{1/p} \leq \|x\| \leq c_2 \left(\sum_k \|(R_{m_{k+1}} - R_{m_k})x\|^p \right)^{1/p}$$

if $x \in X$.
b) Let $X \subset c_0$ *be a closed subspace. Then* X *has the shrinking CBAP if and only if there are a c.a.s.* $\{R_n\}_{n=1}^{\infty}$, *integers* $0 = m_0 < m_1 < m_2 < \ldots$ *and numbers* $c_1 > 0$, $c_2 > 0$ *satisfying*

$$c_1 \sup_k \|(R_{m_{k+1}} - R_{m_k})x\| \leq \|x\| \leq c_2 \sup_k \|(R_{m_{k+1}} - R_{m_k})x\|$$

if $x \in X$.

Proof. Suppose that $X \subset c_0$ and that the inequalities of b) hold. Without loss of generality we can assume that $m_k = k$ for all k (otherwise go over to a suitable subsequence of the R_m). Then X can be identified with the subspace

$$\{\{(R_n - R_{n-1})x\}_{n=1}^\infty \ : \ x \in X\} \ \text{of} \ Y := \left(\sum_n \oplus (R_n - R_{n-1})X\right)_{(c_0)}.$$

So any $\Phi \in X^*$ can be extended to an element $\hat\Phi \in Y^*$ which has the form $\hat\Phi = \{\Phi_n\}_{n=1}^\infty$, $\Phi_n \in ((R_n - R_{n-1})X)^*$ with $\sum_n ||\Phi_n|| < \infty$. Put $\Psi_n = (\Phi_1, \Phi_2, \ldots, \Phi_n, 0, 0, \ldots)$. Then

$$|(\hat\Phi - \hat\Phi \circ R_n)(\{(R_k - R_{k-1})x\}_{k=1}^\infty)|$$

$$\leq \frac{1}{c_1}(1 + \sup_j ||R_j||) \sum_{j>n} ||\Phi_j|| \cdot ||x|| + |(\Psi_n - \Psi_n \circ R_n)(\{(R_k - R_{k-1})x\}_{k=1}^\infty)|$$

$$\leq \frac{1}{c_1}(1 + \sup_j ||R_j||) \sum_{j>n} ||\Phi_j|| \cdot ||x|| + |\Phi_n((R_n^2 - R_{n-1})x)|$$

$$\leq \frac{1}{c_1}(1 + \sup_j ||R_j||) \sum_{j>n} ||\Phi_n|| \cdot ||x|| + 2(\sup_n ||R_n||)^2 ||\Phi_n|| \cdot ||x||.$$

(We used $R_n(R_n - R_{n-1}) = R_n^2 - R_{n-1}$ and $(id - R_n)(R_m - R_{m-1}) = 0$ if $m < n$.) This implies $\lim_n ||\Phi - R_n^*\Phi|| = 0$. Hence X^* has a c.a.s. and X has the shrinking CBAP.

Now assume that $\{R_n\}_{n=1}^\infty$ is a shrinking c.a.s. of X. We prove the case $X \subset l_p$. The case $X \subset c_0$ is similar. Let $p^{-1} + q^{-1} = 1$ and let Φ_n denote the unit vectors in $l_q = l_p^*$. We have, since dim $R_nX < \infty$,

$$\lim_{m\to\infty} ||\Phi_m R_n|| = 0 \ \text{for each} \ n$$

and, since $\{R_n^*\}_{n=1}^\infty$ is a c.a.s. of X^*,

$$\lim_{n\to\infty} ||\Phi_m(id - R_n)|| = 0 \ \text{for each m}.$$

We find, by induction, for any $\epsilon > 0$, integers

$$1 < n_1 < n_2 < \ldots \ \text{and} \ 1 \leq m_1 < m_2 < \ldots \ \text{with}$$

$$||\Phi_j(id - R_{n_k})|| \leq \frac{\epsilon}{2^k}, j = 1, \ldots, m_k \ \text{and} \ ||\Phi_i R_{n_k}|| \leq \frac{\epsilon}{2^k}, i = m_{k+1}, m_{k+1}+1, \ldots$$

Fix j, say $m_l \leq j < m_{l+1}$. Then

$$|\Phi_j(x - (R_{n_{l+1}} - R_{n_{l-1}})x)| \leq \left(\frac{\epsilon}{2^{l+1}} + \frac{\epsilon}{2^{l-1}}\right) ||x||$$

If ϵ is small enough we obtain, using the Minkowski inequality,

$$||x|| \leq c_2 \left(\sum_k ||(R_{m_{k+1}} - R_{m_k})x||^p\right)^{1/p}, \ x \in X.$$

Duality finally yields

$$c_1 \left(\sum_k ||(R_{m_{k+1}} - R_{m_k})x||^p \right)^{1/p} \leq ||x||, \ x \in X \ .$$

(Remember, we considered only $1 < p < \infty$.) For $X \subset c_0$ we use

$$\sup_k ||(R_{m_{k+1}} - R_{m_k})x|| \leq (2 \sup_m ||R_m||) ||x||, \ x \in X \ .$$

□

We finish this section with an l_1-version of 5.2.2.

Theorem 5.2.3 *Let $\{R_n\}_{n=1}^\infty$ be a c.a.s. on X such that, for some constant $c > 0$,*

$$c \sum_k ||(R_{k+1} - R_k)x|| \leq ||x|| \leq \sum_k ||(R_{k+1} - R_k)x||, \quad x \in X.$$

Then X is isomorphic to a dual space and $\{R_n\}_{n=1}^\infty$ is a w^*-shrinking c.a.s.

Proof. Put $E_n = (R_n - R_{n-1})X$ and note that $R_m E_n \subset E_n$ for all m since

$$R_m(R_n - R_{n-1}) = \begin{cases} 0, & m < n-1 \\ R_n - R_{n-1}, & m > n \\ R_n^2 - R_{n-1}, & m = n \\ R_{n-1} - R_{n-1}^2, & m = n - 1 \end{cases} \tag{5.2}$$

Define

$$Y = \{\{e_n^*\}_{n=1}^\infty : e_n^* \in E_n^*,$$

$$(e_{n+1}^* \circ (R_{n+1} - R_n) + e_n^* \circ (R_n - R_{n-1}) + e_{n-1}^* \circ (R_{n-1} - R_{n-2}))|_{E_n}$$

$$= e_n^* \text{ for all } n, \ \lim_n ||e_n^*|| = 0\}$$

$$\subset (\sum_n \oplus E_n^*)_{(c_0)}.$$

(Put $R_{-1} = 0 = R_0$.) For $x \in X$ let $j(x) \in Y^*$ be the element with

$$\langle j(x), \{e_n^*\}_{n=1}^\infty \rangle = \sum_{n=1}^\infty e_n^*((R_n - R_{n-1})x) \ .$$

By assumption this definition makes sense and we obtain $||j(x)|| \leq 1/c||x||$. Consider $x^* \in X^*$ with $x^* = x^* \circ R_m$ for some m. Then, in view of (5.2), $\lim_n ||x^*|_{E_n}|| = 0$. We have

$$((R_{n+1} - R_n) + (R_n - R_{n-1}) + (R_{n-1} - R_{n-2}))(R_n - R_{n-1}) = R_n - R_{n-1} \ .$$

This implies that $\{x^*|_{E_n}\}_{n=1}^\infty \in Y$. Hence $||x|| \le ||j(x)||$ since

$$||x|| = \sup\left\{\left|\sum_{n=1}^\infty x^*((R_n - R_{n-1})x)\right| : x^* \in X^* \text{ with } ||x^*|| \le 1 \text{ and}\right.$$

$$\left. x^* \circ R_m = x^* \text{ for some } m\right\}.$$

Thus $||x|| \le ||j(x)|| \le 1/c||x||$ for all $x \in X$.

We claim $jX = Y^*$. To this end let $y^* \in Y^*$. Then there is an element $\bar{e} = \{e_n\}_{n=1}^\infty \in (\sum \oplus E_n)_{(l_1)}$ with $\bar{e}|_Y = y^*$. Put $x = \sum_{k=1}^\infty e_k$. Then we have, by (1),

$$(R_j - R_{j-1})(x) = \begin{cases} R_1(e_1 + e_2), & j = 1 \\ (R_j - R_{j-1})(e_{j-1} + e_j + e_{j+1}), & j > 1 \end{cases}$$

We obtain, if $\{e_k^*\}_{k=1}^\infty \in Y$, with $e_0 = 0$,

$$\langle j(x), \{e_n^*\}_{n=1}^\infty \rangle = \sum_n e_n^*((R_n - R_{n-1})(e_{n-1} + e_n + e_{n+1}))$$

$$= \sum_n (e_{n+1}^* \circ (R_{n+1} - R_n) + e_n^* \circ (R_n - R_{n-1}) + e_{n-1}^* \circ (R_{n-1} - R_{n-2}))(e_n)$$

$$= \sum_n e_n^*(e_n)$$

by definition of Y. This means $j(x) = y^*$ and so $Y^* = j(X)$. It is clear that the R_n^* define a c.a.s. on Y. $\qquad \square$

Part II

On the Geometry of Müntz Sequences

Now we come to the applications of the first part of our book. Using the concepts and results of Part I we study Müntz sequences. Here we always consider a sequence Λ of real numbers λ_k satisfying

$$\lambda_0 = 0 < \lambda_1 < \lambda_2 < \ldots$$

(In 6.5 we also study negative exponents.)

Let us put $M(\Lambda) = \{t^{\lambda_k}\}_{k=1}^{\infty}$ and let $[M(\Lambda)]_E$ be the closed linear span of $M(\Lambda)$ in a given Banach space E which contains $M(\Lambda)$. We denote the norm in E by $\| \cdot \|_E$. Mainly we consider $E = C = C[0,1]$ and $E = L_p = L_p[0,1]$ for $1 \le p < \infty$. Frequently we drop λ_0 and start with λ_1. In this context we also consider $E = C_0 = C_0[0,1]$, the subspace of $C[0,1]$ consisting of all continuous functions which vanish at 0.

We start with Chap. 6 where we deal with the classical versions of the Müntz theorem i.e. with cases where $[M(\Lambda)]_E \ne E$. This gives rise to a study of the Banach space $[M(\Lambda)]_E$. After a preparatory analysis of Müntz polynomials $f(t) = \sum_{k=1}^{n} \alpha_k t^{\lambda_k}$ in Chaps. 7 and 8 we discuss bounded approximation properties in $[M((\Lambda)]_E$ for $E = C$ or $E = L_p$ in Chap. 9. Here we give a complete isomorphic classification of $[M(\Lambda)]_E$ for quasilacunary Λ.

In Chap. 10 we show, among other things, that there are at least two different ismorphism classes and infinitely many isometry classes of $[M(\Lambda)]_C$. Chapter 11 deals with generalizations of Müntz spaces while, finally, in Chap. 12 we study limiting norms on finite dimensional Müntz spaces.

6

Coefficient Estimates and the Müntz Theorem

Fix $\Lambda = \{\lambda_k\}_{k=1}^{\infty}$ satisfying $\lambda_0 = 0 < \lambda_1 < \lambda_2 < \dots$. The classical Müntz theorem states that $[M(\Lambda)]_E \neq E$ if and only if $\sum_{k=1}^{\infty} \frac{1}{\lambda_k} < \infty$ where $E = C_0$ or $E = L_p$, $1 \leq p < \infty$. We start with a proof for this theorem which essentially characterizes when $M(\Lambda)$ is a minimal sequence. This is the case whenever we have "good" coefficient estimates for the Müntz polynomials $\sum_{k=1}^{n} \alpha_k t^{\lambda_k}$. Instead of only considering C_0 or L_p we shall work in a much more general setting.

In the following let E be the completion of the vector space C_0 with respect to a given seminorm $||\cdot||_E$ with $||t^{\lambda_k}||_E \neq 0$ for all $\lambda_k \in \Lambda$. We assume further that the operators

$$T_\rho : [M(\Lambda)]_E \rightarrow [M(\Lambda)]_E \quad \text{with} \quad (T_\rho f)(t) = f(\rho t), \quad t \in [0,1] \,,$$

for all $f \in \text{span } M(\Lambda)$, are bounded with respect to $||\cdot||_E$ whenever $0 < \rho < 1$ and that

$$\sup_{\rho_0 \leq \rho \leq 1} ||T_\rho|| < \infty \quad \text{for all} \quad \rho_0 \in \,]0,1[\tag{6.1}$$

where $||T_\rho||$ is the corresponding operator norm.

Examples for $||\cdot||_E$ include the sup-norm and the L_p-norms on $[0,1]$. More generally, we can consider Orlicz function spaces over $[0,1]$ (see [84]). Here let $M : [0,\infty[\rightarrow [0,\infty[$ be a continuous convex increasing function with $M(0) = 0$ and $\lim_{t\to\infty} M(t) = \infty$ and put

$$||f|| = \inf \left\{ \delta > 0 \; : \; \int_0^1 M\left(\frac{|f(t)|}{\delta}\right) < 1 \right\}$$

Also, occasionally, we apply some of the following results to suitable quotient spaces.

6.1 The Müntz Theorem

We begin with a technical lemma.

Lemma 6.1.1 *For arbitrary real numbers* a_i, b_i, $i = 1, \ldots n$, *we have*

$$\prod_{i=1}^{n}\prod_{j=1}^{n}(a_i + b_j)\begin{vmatrix}\frac{1}{a_1+b_1} & \cdots & \frac{1}{a_1+b_n} \\ \vdots & & \vdots \\ \frac{1}{a_n+b_1} & \cdots & \frac{1}{a_n+b_n}\end{vmatrix} = \prod_{i=1}^{n}\prod_{\substack{j=1 \\ j<i}}^{n}(a_i - a_j)(b_i - b_j)$$

Proof. We may assume that $a_1, \ldots, a_n, b_1, \ldots, b_n$ are mutually different. Let $l = l(a_1, \ldots, b_n)$ and $r = r(a_1, \ldots, b_n)$ be the left-hand side and the right-hand side, resp., of the above equation. l and r are polynomials in a_1 of degree $n-1$ with zeros $a_2, \ldots, b_n$. (Here we keep $a_2, \ldots, b_n$ fixed.) Hence $l = f_1 r$ for some constant $f_1 = f_1(a_2, \ldots, b_n)$. Next, we regard l and r as polynomials in a_2 and obtain similarly $l = f_2 r$ for some constant $f_2 = f_2(a_1, a_3, \ldots, b_n)$. This means $f_1 = f_2$ and therefore f_1 is independent of a_1 and a_2. Continuation of this argument eventually yields that f_1 is independent of $a_1, \ldots, b_n$.

Taking the limits $b_1 \to -a_1$, ..., $b_n \to -a_n$ we see that r becomes

$$\prod_{j<i}(a_i - a_j)(-a_i + a_j) = \prod_{i=1}^{n}\prod_{\substack{j=1 \\ i\neq j}}^{n}(a_i - a_j) \, .$$

Moreover,

$$l = \prod_{i=1}^{n}\prod_{\substack{j=1 \\ i\neq j}}^{n}(a_i + b_j)\begin{vmatrix} 1 & \frac{a_1+b_1}{a_1+b_2} & \cdots & \frac{a_1+b_1}{a_1+b_n} \\ \frac{a_2+b_2}{a_2+b_1} & 1 & \cdots & \frac{a_2+b_2}{a_2+b_n} \\ \vdots & & & \vdots \\ \frac{a_n+b_n}{a_n+b_1} & \cdots & \cdots & 1 \end{vmatrix}$$

changes to

$$\prod_{i=1}^{n}\prod_{\substack{j=1 \\ i\neq j}}^{n}(a_i - a_j)\begin{vmatrix} 1 & 0 & 0 & \ldots & 0 \\ 0 & 1 & 0 & \ldots & 0 \\ \vdots & & \ddots & & \vdots \\ 0 & \ldots & 0 & 1 & 0 \\ 0 & \ldots & 0 & 0 & 1 \end{vmatrix} = \lim r \, .$$

Since $\lim l = f_1 \lim r$ we see that $f_1 = 1$.   □

At first we give a coefficient estimate for the sup-norm $\|\cdot\|_C$ in $C = C[0,1]$.

Proposition 6.1.2 *For any Müntz polynomial* $f(t) = \sum_{k=1}^{n} \alpha_k t^{\lambda_k}$ *and any* m *we have*

$$|\alpha_m| \le \left(\sum_{j=1}^{n}\frac{1}{\lambda_j + 1}\right)\left(\frac{\prod_{i=1}^{n}(\lambda_i + \lambda_m + 1)^2}{\prod_{i=1;\ i\neq m}^{n}(\lambda_i - \lambda_m)^2}\right)\|f\|_C$$

Proof. We want to find functions $g_m(t) = \sum_{l=1}^{n} \gamma_{m,l} t^{\lambda_l} \in L_1[0,1]$ such that

$$\int_0^1 g_m(t) t^{\lambda_j} dt = \delta_{m,j}, \quad j = 1, \ldots, n, \tag{6.2}$$

where $\delta_{m,j} = \begin{cases} 1, & m = j \\ 0, & m \neq j \end{cases}$. Hence we have to find $\gamma_{m,l}$ such that

$$\sum_{l=1}^{n} \gamma_{m,l} \frac{1}{\lambda_l + \lambda_j + 1} = \delta_{m,j}, \quad j = 1, \ldots, n.$$

Let $\det(b_1, \ldots, b_n)$ be the determinant with the columns $b_1, \ldots, b_n$. Putting $x_m = (\delta_{m,1}, \ldots, \delta_{m,n})^t$ and $a_l = ((\lambda_1 + \lambda_l + 1)^{-1}, \ldots, (\lambda_n + \lambda_l + 1)^{-1})^t$, we see that Cramer's rule implies

$$\gamma_{m,l} = \frac{\det(a_1, \ldots, a_{l-1}, x_m, a_{l+1}, \ldots, a_n)}{\det(a_1, \ldots, a_n)}.$$

Hence with

$$a_l' = \left(\frac{1}{\lambda_1 + \lambda_l + 1}, \ldots, \frac{1}{\lambda_{m-1} + \lambda_l + 1}, \frac{1}{\lambda_{m+1} + \lambda_l + 1}, \ldots, \frac{1}{\lambda_n + \lambda_l + 1} \right)^t$$

we obtain

$$\gamma_{m,l} = (-1)^{l+m} \frac{\det(a_1', \ldots, a_{l-1}', a_{l+1}', \ldots, a_n')}{\det(a_1, \ldots, a_n)}.$$

Lemma 6.1.1 helps us to estimate $|\gamma_{m,l}|$:

$$|\gamma_{m,l}| \leq \left(\frac{\prod_{j<i;\ i,j \neq m} (\lambda_i - \lambda_j)^2}{\prod_{i,j \neq m} (\lambda_i + \lambda_j + 1)} \right) \left(\frac{\prod_{j<i} (\lambda_i - \lambda_j)^2}{\prod_{i,j} (\lambda_i + \lambda_j + 1)} \right)^{-1}$$

$$= \frac{\prod_{i=1}^{n} (\lambda_i + \lambda_m + 1)^2}{\prod_{i=1;\ i \neq m}^{n} (\lambda_i - \lambda_m)^2}$$

We have

$$\|g_m\|_{L_1} = \int_0^1 |g_m(t)| dt$$

$$\leq \sum_{l=1}^{n} \frac{1}{\lambda_l + 1} |\gamma_{m,l}|$$

$$\leq \left(\sum_{l=1}^{n} \frac{1}{\lambda_l + 1} \right) \frac{\prod_{i=1}^{n} (\lambda_i + \lambda_m + 1)^2}{\prod_{i=1;\ i \neq m}^{n} (\lambda_i - \lambda_m)^2}$$

Now, (6.2) yields

$$|\alpha_m| = \left| \int_0^1 g_m(t) f(t) dt \right| \leq \|g_m\|_{L_1} \cdot \|f\|_C$$

which implies the proposition. $\qquad\qquad\qquad\qquad\qquad\qquad\qquad\qquad\square$

From Proposition 6.1.2 we easily derive

Corollary 6.1.3 *Let $\sum_{j=1}^{\infty} 1/\lambda_j < \infty$. Then $M(\Lambda) = \{t^{\lambda_j}\}_{j=1}^{\infty}$ is a minimal system with respect to the sup-norm in $C[0,1]$. Moreover, there is a constant $d > 0$ such that*

$$|f(t)| \leq dt^{\lambda_1}\|f\|_C, \quad 0 \leq t \leq 1,$$

for any $f \in [M(\Lambda)]_C$. In particular,

$$\|f\|_C = \sup\{|f(s)| : d^{-1/\lambda_1} \leq s \leq 1\}.$$

Proof. With $f(t) = \sum_k \alpha_k t^{\lambda_k}$ Proposition 6.1.2 implies

$$|\alpha_m| \leq$$

$$\left(\sum_{j=1}^{\infty} \frac{1}{\lambda_j + 1}\right)(2\lambda_m + 1)^2 \prod_{i>m}\left(1 + \frac{2\lambda_m + 1}{\lambda_i - \lambda_m}\right) \prod_{i<m}\left(1 + \frac{2\lambda_i + 1}{\lambda_m - \lambda_i}\right)\|f\|_C \leq$$

$$\left(\sum_{j=1}^{\infty} \frac{1}{\lambda_j + 1}\right)(2\lambda_m + 1)^2 \exp\left(2\sum_{i>m}\frac{2\lambda_m + 1}{\lambda_i - \lambda_m} + 2\sum_{i<m}\frac{2\lambda_i + 1}{\lambda_m - \lambda_i}\right)\|f\|_C < \infty$$

Hence the coefficient functionals for $\{t^{\lambda_k}\}_{k=1}^{\infty}$ are bounded which proves the first statement of the corollary (see 2.2.2).

For each fixed $t > 0$ the functional Φ_t with $\Phi_t(f) = f(t)/t^{\lambda_1}$, $f \in [M(\Lambda)]_C$, is bounded on $[M(\Lambda)]_C$. By the preceding argument, also Φ_0 with $\Phi_0(f) = \lim_{t \to 0} f(t)/t^{\lambda_1} = \alpha_1$ is bounded on $[M(\Lambda)]_C$. Since $\sup_t |\Phi_t(f)| < \infty$ for each $f \in [M(\Lambda)]_C$ the uniform boundedness principle yields $d = \sup_t \|\Phi_t\| < \infty$. This proves $|f(t)| \leq dt^{\lambda_1}\|f\|_C$ for any $t \in [0,1]$ and any $f \in [M(\lambda)]_C$. In particular, $|f(s)| < \|f\|_C$ whenever $0 \leq s < d^{-1/\lambda_1}$ which yields the remaining part of the corollary. $\qquad\square$

Now we turn to general norms. As already indicated we consider an arbitrary (semi-)norm on a space E which contains $C_0[0,1]$ such that $\|t^{\lambda_k}\|_E \neq 0$ for all $\lambda_k \in \Lambda$. Moreover, we assume that $\sup_{\rho_0 \leq \rho \leq 1}\|T_\rho\| < \infty$ for any $\rho_0 > 0$ where $T_\rho : [M(\Lambda)]_E \to [M(\Lambda)]_E$ is the operator with $(T_\rho f)(t) = f(\rho t)$, $t \in [0,1]$, for $f \in \text{span } M(\Lambda)$.

As another consequence we obtain one direction of the Müntz theorem in this general setting.

Proposition 6.1.4 *Assume that $\sum_{k=1}^{\infty} 1/\lambda_k < \infty$. Then $\{t^{\lambda_k}\}_{k=1}^{\infty}$ is a minimal system in $[M(\Lambda)]_E$.*

Actually, for any Müntz polynomial $f(t) = \sum_k \alpha_k t^{\lambda_k}$ and any index m we have

$$\|\alpha_m t^{\lambda_m}\|_E \leq \left(\sum_{j=1}^{\infty} \frac{1}{\lambda_j + 1}\right) \frac{\prod_{i=1}^{\infty}(\lambda_i + \lambda_m + 1)^2}{\prod_{i=1;\ i \neq m}^{\infty}(\lambda_i - \lambda_m)^2}\left(\sup_{d^{-1/\lambda_1} \leq \rho \leq 1}\|T_\rho\|\right)\|f\|_E$$

where d is the constant of Corollary 6.1.3

Proof. Fix a functional $\Phi \in E^*$ of norm one. For any $f(t) = \sum_{k=1}^n \alpha_k t^{\lambda_k}$ put

$$f_\Phi(\rho) = \Phi(T_\rho f) = \sum_{k=1}^n \alpha_k \Phi(t^{\lambda_k}) \rho^{\lambda_k} .$$

Then 6.1.3 yields, for any m, a constant $d_m > 0$ with

$$|\alpha_m \Phi(t^{\lambda_m})| \le d_m \sup_{0 \le \rho \le 1} |f_\Phi(\rho)| = d_m \sup_{d^{-1/\lambda_1} \le \rho \le 1} |f_\Phi(\rho)|$$

$$\le d_m (\sup_{d^{-1/\lambda_1} \le \rho \le 1} ||T_\rho||) ||f||_E$$

where d is the constant of Corollary 6.1.3. Hence

$$|\alpha_m| \cdot ||t^{\lambda_m}||_E \le d_m (\sup_{d^{-1/\lambda_1} \le \rho \le 1} ||T_\rho||) ||f||_E$$

which implies that $\{t^{\lambda_k}\}_{k=1}^\infty$ is minimal in $[M(\lambda)]_E$. According to 6.1.1, we have

$$d_m = \left(\sum_{j=1}^\infty \frac{1}{\lambda_j + 1} \right) \frac{\prod_{i=1}^\infty (\lambda_i + \lambda_m + 1)^2}{\prod_{i=1;\ i \ne m}^\infty (\lambda_i - \lambda_m)^2}$$

which is bounded if $\sum_{j=1}^\infty 1/\lambda_j < \infty$. $\square$

For the other direction we use the elegant argument of von Golitschek [33].

Theorem 6.1.5 *Assume that, for some constant $\kappa > 0$, $||f||_E \le \kappa ||f||_C$ whenever $f \in C_0$. Moreover suppose that there is $\mu > 0$, $\mu \notin \Lambda$, with $||t^\mu||_E \ne 0$. Then the following are equivalent:*
(i) $\sum_{k=1}^\infty 1/\lambda_k < \infty$
(ii) $[M(\Lambda)]_E \ne E$
(iii) $\{t^{\lambda_k}\}_{k=1}^\infty$ *is a minimal system in* $[M(\Lambda)]_E$.

Proof. (ii) $\Rightarrow$ (i): (von Golitschek, [33]) Assume $\sum_k 1/\lambda_k = \infty$. Let $m \ne \lambda_k$, $k = 1, 2, \ldots$. Define $q_0(t) = t^m$ and

$$q_n(t) = (\lambda_n - m) t^{\lambda_n} \int_t^1 q_{n-1}(s) s^{-1-\lambda_n} ds, \quad n = 1, 2, \ldots .$$

Then q_n has the form

$$q_n(t) = \theta t^m - \sum_{k=0}^n a_{n,k} t^{\lambda_k} \quad \text{for some } a_{n,k}, \quad \theta \in \{1, -1\}$$

which can be easily seen by induction. We obtain $||q_0||_C = 1$ and, by construction,
$||q_n||_C \le |1 - m/\lambda_n| \cdot ||q_{n-1}||_C$. By assumption, this implies

$$||q_n||_E \leq \kappa ||q_n||_C \leq \prod_{k=0}^{n} \left| 1 - \frac{m}{\lambda_k} \right| \to 0 \text{ as } n \to \infty .$$

Hence $t^m \in [M(\Lambda)]_E$ for all m and therefore $[M(\Lambda)]_E = E$. This contradicts (ii).

(i) $\Rightarrow$ (ii): Take $\mu > 0$ with $\mu \notin \Lambda$. Now, 6.1.4 applied to $\tilde{\Lambda} = \{\mu, \lambda_1, \lambda_2, \ldots\}$ shows that $t^\mu \notin [M(\Lambda)]_E$ which implies (ii).

(i) $\Rightarrow$ (iii): This follows from 6.1.4.

(iii) $\Rightarrow$ (i): Assume $\sum_k 1/\lambda_k = \infty$. Put $\tilde{\Lambda} = \{\lambda_2, \lambda_3, \ldots\}$. Then also $\sum_{k \geq 2} 1/\lambda_k = \infty$ and hence, by what we have proved already, $[M(\tilde{\Lambda})]_E = E$. This implies $t^{\lambda_1} \in$ closed span $\{t^{\lambda_k} : k = 2, 3, \ldots\}$, a contradiction to minimality. $\qquad\square$

As mentioned before 6.1.5 includes the cases $E = L_p$ and $E = C_0$.

For $E = L_p$, $1 \leq p < \infty$, we even could consider sequences Λ with $-1/p < \lambda_1 < \lambda_2 < \ldots$. 6.1.5 would remain valid if we replaced (i) by

$$\sum_{k=1}^{\infty} \frac{\lambda_k + 1/p}{(\lambda_k + 1/p)^2 + 1} < \infty .$$

See [12] for details.

6.1.5 tells us that the case $\sum_{k=1}^{\infty} 1/\lambda_k = \infty$ is not too exciting. However, if $\sum_{k=1}^{\infty} 1/\lambda_k < \infty$ we obtain new Banach spaces $[M(\Lambda)]_E$. Therefore, in the following we will be mostly concerned with the latter case. It turns out that further coefficient theorems are essential for the investigation.

6.2 The Clarkson-Erdös Theorem

Here we want to refine the estimate of Proposition 6.1.4 provided that the *gap condition*

$$\inf_k (\lambda_{k+1} - \lambda_k) > 0$$

is satisfied.

Lemma 6.2.1 *Assume* $\sum_{k=1}^{\infty} 1/\lambda_k < \infty$ *and* $c := \inf_k (\lambda_{k+1} - \lambda_k) > 0$. *Put* $L_m = \sup\{j : \lambda_j \leq 2\lambda_m\}$. *Then, for any Müntz polynomial* $f(t) = \sum_k \alpha_k t^{\lambda_k} \in [M(\Lambda)]_E$ *we have*

$$||\alpha_m t^{\lambda_m}||_E \leq \left(\sum_{j=1}^{\infty} \frac{1}{\lambda_j + 1} \right) (2\lambda_m + 1)^2 \left(\frac{1}{c} \right)^{2L_m - 2} \left(\frac{(2\lambda_m + 1)^{m-1}}{(m-1)!} \right)^2 .$$

$$\left(\frac{(3\lambda_m + 1)^{L_m - m}}{(L_m - m)!} \right)^2 \exp \left(4 \sum_{i \geq L_m} \frac{2\lambda_m + 1}{\lambda_i} \right) \left(\sup_{d^{-1/\lambda_1} \leq \rho \leq 1} ||T_\rho|| \right) ||f||_E$$

where d *is the constant of 6.1.3.*

Proof. According to 6.1.4 we have

$$\|\alpha_m t^{\lambda_m}\|_E \le \left(\sum_{j=1}^{\infty} \frac{1}{\lambda_j + 1}\right) (2\lambda_m + 1)^2 \prod_{i<m} \left(\frac{\lambda_i + \lambda_m + 1}{\lambda_m - \lambda_i}\right)^2 \cdot$$

$$\prod_{m<i<L_m} \left(\frac{\lambda_i + \lambda_m + 1}{\lambda_i - \lambda_m}\right)^2 \exp\left(2 \sum_{i \ge L_m} \frac{2\lambda_m + 1}{\lambda_i - \lambda_m}\right) (\sup_{d^{-1/\lambda_1} \le \rho \le 1} \|T_\rho\|) \|f\|_E$$

The gap condition yields $\lambda_i - \lambda_m \ge (i - m)c$ if $i > m$ and $\lambda_m - \lambda_i \ge (m - i)c$ if $m > i$. Hence

$$\|\alpha_m t^{\lambda_m}\|_E \le \left(\sum_{j=1}^{\infty} \frac{1}{\lambda_j + 1}\right) (2\lambda_m + 1)^2 \left(\frac{1}{c}\right)^{2L_m - 2} \left(\frac{(2\lambda_m + 1)^{m-1}}{(m - 1)!}\right)^2 \cdot$$

$$\left(\frac{(3\lambda_m + 1)^{L_m - m}}{(L_m - m)!}\right)^2 \exp\left(4 \sum_{i \ge L_m} \frac{2\lambda_m + 1}{\lambda_i}\right) (\sup_{d^{-1/\lambda_1} \le \rho \le 1} \|T_\rho\|) \|f\|_E$$

$\square$

Of course the lemma remains true if Λ is finite. Now we prove the Clarkson-Erdös Theorem [19].

Theorem 6.2.2 *Let $\sum_{j=1}^{\infty} 1/\lambda_j < \infty$ and assume $\inf_j(\lambda_{j+1} - \lambda_j) > 0$. Then, for any $\epsilon > 0$, there is a constant $M > 0$ such that*

$$\|\alpha_m t^{\lambda_m}\|_E \le (1 + \epsilon)^{\lambda_m} \|f\|_E, \quad if \quad m \ge M,$$

for all Müntz polynomials $f(t) = \sum_k \alpha_k t^{\lambda_k}$.
Moreover, for the constants L_m of 6.2.1 we have $\lim_{m\to\infty} L_m/\lambda_m = 0$.

Proof. Using the Stirling formula $n! \ge n^n e^{-n}$ for any positive integer n we conclude with 6.2.1 that

$$\|\alpha_m t^{\lambda_m}\|_E \le \left(\sum_{j=1}^{\infty} \frac{1}{\lambda_j + 1}\right) (2\lambda_m + 1)^2 \left(\frac{1}{c}\right)^{2L_m - 2} \left(\frac{2\lambda_m + 1}{m - 1}\right)^{2m - 2} \cdot$$

$$\left(\frac{3\lambda_m + 1}{L_m - m}\right)^{2L_m - 2m} e^{2L_m - 2} \exp\left(4 \sum_{i \ge L_m} \frac{2\lambda_m + 1}{\lambda_i}\right) (\sup_{d^{-1/\lambda_1} \le \rho \le 1} \|T_\rho\|) \|f\|_E$$

where $c = \inf_j(\lambda_{j+1} - \lambda_j)$.

We claim that $\lim_{m\to\infty} L_m/\lambda_m = 0$. Indeed, fix $\epsilon > 0$ and take m_0 such that $\sum_{k \ge m_0} 1/\lambda_k \le \epsilon/3$. Let $m_1 \ge m_0$ be such that $m_0/\lambda_{m_1} \le \epsilon/3$. Then, for any $m \ge m_1$, we obtain

$$\frac{L_m}{\lambda_m} = \frac{m_0}{\lambda_m} + \frac{m - m_0}{\lambda_m} \le \frac{\epsilon}{3} + \sum_{j=m_0+1}^{m} \frac{1}{\lambda_j} + \sum_{j=m+1}^{L_m} \frac{1}{\lambda_j} \le \epsilon \, .$$

Now, using $\lim_{x \to 0} x^{-x} = 1$ we see that

$$\lim_{m \to \infty} \left(\sum_{j=1}^{\infty} \frac{1}{\lambda_j + 1} \right)^{1/\lambda_m} (2\lambda_m + 1)^{2/\lambda_m} \left(\frac{1}{c} \right)^{(2L_m - 2)/\lambda_m} \left(\frac{2\lambda_m + 1}{m - 1} \right)^{(2m-2)/\lambda_m} \, .$$

$$\left(\frac{3\lambda_m + 1}{L_m - m} \right)^{(2L_m - 2m)/\lambda_m} e^{(2L_m - 2)/\lambda_m} \exp \left(4 \sum_{i \ge L_m} \frac{2 + 1/\lambda_m}{\lambda_i} \right) \, .$$

$$\left(\sup_{d^{-1/\lambda_1} \le \rho \le 1} \|T_\rho\| \right)^{1/\lambda_m} = 1$$

which proves the theorem. $\qquad\square$

Of course, the assumptions of Theorem 6.2.2 are fulfilled in particular for the sup-norm and the L_p-norms on $[0, 1]$.

Theorem 6.2.2 has important consequences for the elements in $[M(\Lambda))]_E$.

Theorem 6.2.3 *Suppose that* $\liminf_{n \to \infty} \|t^{\lambda_n}\|_E^{1/\lambda_n} \ge 1$. *Moreover, let* $\sum_{j=1}^{\infty} 1/\lambda_j < \infty$ *and* $\inf_j (\lambda_{j+1} - \lambda_j) > 0$. *Then, for any* $f \in [M(\Lambda)]_E$ *there are suitable* $\alpha_k \in \mathbf{R}$ *such that*

$$f(t) = \sum_{k=1}^{\infty} \alpha_k t^{\lambda_k}, \quad t \in [0, 1[\, ,$$

where the series converges uniformly on all compact subsets of $[0, 1[$. *In particular, any* $f \in [M(\Lambda)]_E$ *is real-analytic on* $]0, 1[$. *If* Λ *consists of integers then any* $f \in [M(\Lambda)]_E$ *is real-analytic on* $[0, 1[$.

Proof. If $f \in [M(\Lambda)]_E$ there are $f_n(t) = \sum_{k=1}^{m_n} \alpha_{n,k} t^{\lambda_k}$ converging to f with respect to $\| \cdot \|_E$. Since $\|t^{\lambda_k}\|_E \neq 0$, in view of 6.1.4, this implies that $\alpha_k := \lim_{n \to \infty} \alpha_{n,k}$ exists for all k. We infer from 6.2.2 that, for every $\epsilon > 0$, there is $M > 0$ with

$$|\alpha_k| \cdot \|t^{\lambda_k}\|_E \le (1 + \epsilon)^{\lambda_k} \|f\|_E \quad \text{if} \quad k \ge M$$

Since $\liminf_{k \to \infty} \|t^{\lambda_k}\|_E^{1/\lambda_k} \ge 1$ we obtain $\limsup_{k \to \infty} |\alpha_k|^{1/\lambda_k} \le 1$. Hence $\sum_{k=1}^{\infty} \alpha_k t^{\lambda_k}$ converges uniformly on all compact subsets of $[0, 1[$. Moreover, $\sum_{k=1}^{\infty} \alpha_k t^{\lambda_k}$ converges uniformly on compact subsets of $\{z \in \mathbf{C} : |z| < 1, \mathrm{Re}\, z > 0\}$. The functions z^{λ_k} are analytic on $\{z \in \mathbf{C} : \mathrm{Re}\, z > 0\}$. This implies that f is real-analytic on $]0, 1[$ and even on $[0, 1[$ if Λ is an integer sequence. $\qquad\square$

It is remarkable that not only span $M(\Lambda)$ but even its completion $[M(\Lambda)]_E$ consists exclusively of real-analytic functions. Again, 6.2.3 can be applied to the special case of $E = C_0$ or $E = L_p$, $1 \le p < \infty$. The converse of the first part of Theorem 6.2.3 is valid under more restrictive assumptions. We state this as a corollary only for $E = C_0$ and $E = L_p$.

Corollary 6.2.4 *Assume that $\sum_{j=1}^{\infty} 1/\lambda_j < \infty$ and $\inf_j(\lambda_{j+1} - \lambda_j) > 0$. Let $E = L_p$, $1 \le p < \infty$, or $E = C_0$. Then $f \in [M(\Lambda)]_E$ if and only if $f \in E$ and there are suitable $\alpha_k \in \mathbf{R}$ such that*

$$f(t) = \sum_{k=1}^{\infty} \alpha_k t^{\lambda_k}, t \in [0,1[,$$

where the series converges uniformly on all compact subsets of $[0,1[$.

Proof. In view of 6.2.3 it remains to prove the only-if part. For any $\rho \in]0,1[$ the operator T_ρ can be extended to E by defining $(T_\rho g)(t) = g(\rho t)$, $t \in [0,1]$, whenever $g \in E$. For our special E we have $\lim_{\rho \to 1} \|T_\rho g - g\|_E = 0$ whenever $g \in E$. Now consider $f \in E$ such that $f(t) = \sum_{k=1}^{\infty} \alpha_k t^{\lambda_k}$ for some α_k and the series converges compactly on $[0,1[$. Then $\limsup_{k \to \infty} |\alpha_k|^{1/\lambda_k} \le 1$. For any $\epsilon > 0$ there is $\rho < 1$ with $\|T_\rho f - f\|_E < \epsilon$. Put $f_n(t) = \sum_{k=1}^{n} \alpha_k t^{\lambda_k}$. Then $T_\rho f_n \in [M(\Lambda)]_E$ for all n and $\lim_{n \to \infty} \|T_\rho f_n - T_\rho f\|_E = 0$ (because we have $\|T_\rho(f_n - f)\|_E \le \sum_{k=n+1}^{\infty} |\alpha_k| \rho^{\lambda_k}$). We infer $f = \lim_{\rho \to 1} T_\rho f \in [M(\Lambda)]_E$. □

For further applications of 6.2.3 see Sect. 8.4.

If we consider $\{\lambda_k\}_{k=0}^{\infty}$ (with $\lambda_0 = 0$) instead of $\Lambda = \{\lambda_k\}_{k=1}^{\infty}$ then all the results corresponding to the assertions of Sects. 6.1 and 6.2 remain true (with virtually the same proofs). Instead of $E = C_0[0,1]$ we take here $E = C[0,1]$.

6.3 Lacunary and Quasilacunary Müntz Sequences

In the preceding sections we saw that $M(\Lambda)$ is a minimal system if and only if $\sum_{k=1}^{\infty} 1/\lambda_k < \infty$. However, $M(\Lambda)$ is not necessarily uniformly minimal or a basis. For example, if we take the sup-norm $\| \cdot \|_C$ on $[0,1]$ then we easily see that $\|t^\lambda - t^\mu\|_C \le (\mu - \lambda)/\mu$ whenever $0 < \lambda < \mu$ (see 7.3.1). Hence if $\Lambda = \{\lambda_k\}_{k=1}^{\infty}$ and $\inf_k \lambda_{k+1}/\lambda_k = 1$ then $M(\Lambda)$ need not be even separated. (For the definition see 2.2.1.)

However, in the special case of lacunary Λ the situation is different. This will be treated in full detail in 9.2. Here we show, as a first step, that for lacunary Λ we obtain much better coefficient estimates.

Definition 6.3.1 $\Lambda = \{\lambda_k\}_{k=1}^{\infty}$ *will be called*
    (1) lacunary if $\lambda_{k+1}/\lambda_k \ge q$, $k = 1, 2, \ldots$, for some $q > 1$,
    (2) quasilacunary, if for some increasing sequence $\bar{n} = \{n_k\}_{k=1}^{\infty}$ of integers and some $q > 1$ we have $\lambda_{n_k+1}/\lambda_{n_k} \ge q$, $k = 1, 2, \ldots$, and $\sup_k(n_{k+1} - n_k) < \infty$.

For example, $\Lambda = \{2^k\}_{k=1}^{\infty}$ is lacunary while $\{2^k\}_{k=1}^{\infty} \cup \{2^k + 1\}_{k=1}^{\infty}$ is quasi-lacunary but not lacunary.

We give a more comprehensive classification of the different types of sequences Λ in 7.1. Here we want to discuss some direct applications of the results in 6.1 and 6.2.

Again let E satisfy the assumptions of the beginning of this chapter.

Consider the function $\eta(y) = 4y(1 - y)$. We have $0 \leq \eta(y) \leq 1$ for all $y \in [0, 1]$ and $\eta(y) = 1$ if and only if $y = 1/2$.

Proposition 6.3.2 *Asssume that $\Lambda = \{\lambda_k\}_{k=1}^{\infty}$ is lacunary. Then there is an integer $M > 0$, independent of n, with*

$$\sum_{\substack{j=1 \\ j \neq n}}^{\infty} \eta(2^{-\lambda_j/\lambda_n})^M \leq 1/2 \ .$$

Furthermore, for any Müntz polynomial $f(t) = \sum_k \alpha_k t^{\lambda_k} \in [M(\Lambda)]_E$ we have

$$\|\alpha_n t^{\lambda_n}\|_E \leq 4^M \sup_{k=1,2,\ldots} (\|T_{2^{-1/\lambda_k}}\| + \|T_{2^{-1/\lambda_k}}\|^2)^M \|f\|_E$$

Proof. We use a trick due to Ingham (see [58], Theorem 115).

Assume that $\inf_k \lambda_{k+1}/\lambda_k = q > 1$. Let n be such that $\|\alpha_n t^{\lambda_n}\|_E = max_k \|\alpha_k t^{\lambda_k}\|_E$. Consider the function

$$\eta(2^{-\lambda/\lambda_n}) = 4 \cdot 2^{-\lambda/\lambda_n}(1 - 2^{-\lambda/\lambda_n})$$

We have

$$-\frac{\lambda_j}{\lambda_n} \leq -q^{j-n} \text{ if } j > n \text{ and } -\frac{\lambda_j}{\lambda_n} \geq -\left(\frac{1}{q}\right)^{n-j} \text{ if } j < n \ .$$

This implies, since $\eta(y)$ is increasing if $0 < y < 1/2$ and decreasing if $1/2 < y < 1$,

$$\eta(2^{-\lambda_j/\lambda_n}) \leq \eta(2^{-1/q^{n-j}}) = 4 \cdot 2^{-1/q^{n-j}}(1 - 2^{-1/q^{n-j}}) \leq \frac{4}{q^{n-j}} \text{ if } j < n$$

and

$$\eta(2^{-\lambda_j/\lambda_n}) \leq \eta(2^{-q^{j-n}}) \leq 4 \cdot 2^{-q^{j-n}} \text{ if } j > n \ .$$

Taking into account that $0 \leq \eta(2^{-\lambda_j/\lambda_n}) < 1$ we find M depending only on q and not on n such that

$$\sum_{j \neq n}^{\infty} \eta(2^{-\lambda_j/\lambda_n})^M \leq \sum_{k=1}^{\infty} \eta(2^{-1/q^k})^M + \sum_{k=1}^{\infty} \eta(2^{-q^k})^M \leq \frac{1}{2}$$

Put $S = (4T_{2^{-1/\lambda_n}} - 4T_{2^{-2/\lambda_n}})^M$. Then we obtain

$$(Sf)(t) = \sum_j \eta(2^{-\lambda_j/\lambda_n})^M \alpha_j t^{\lambda_j} .$$

Since $\eta(2^{-\lambda_j/\lambda_n})^M = 1$ if $j = n$ we conclude

$$\frac{1}{2}\|\alpha_n t^{\lambda_n}\|_E \le \|\alpha_n t^{\lambda_n}\|_E - \|\sum_{j \ne n} \eta(2^{-\lambda_j/\lambda_n})^M \alpha_j t^{\lambda_j}\|_E$$

$$\le \|Sf\|_E$$

$$\le 4^M (\|T_{2^{-1/\lambda_n}}\| + \|T_{2^{-1/\lambda_n}}\|^2)^M \|f\|_E$$

$$\le 4^M \sup_{k=1,2,\ldots} (\|T_{2^{-1/\lambda_k}}\| + \|T_{2^{-1/\lambda_k}}\|^2)^M \|f\|_E$$

which implies 6.3.2   □

According to 2.2, the preceding proposition implies that, if Λ is lacunary, then $M(\Lambda)$ is uniformly minimal. Namely, it is δ-minimal with

$$\delta = 4^{-M} \inf\{\|T_{2^{-1/\lambda_n}}\|^M (1 + \|T_{2^{-1/\lambda_n}}\|)^M : n = 1, 2 \ldots\}$$

where M is such that

$$\sum_{k=1}^\infty \left(\eta\left(\frac{1}{2^{1/q^k}}\right)^M + \eta\left(\frac{1}{2^{q^k}}\right)^M\right) \le \frac{1}{2} .$$

This means in particular that $M(\Lambda)$ is uniformly minimal.

If $E = L_p$, $1 \le p < \infty$ or $E = C_0$, we even have that $M(\Lambda)$ is uniformly minimal if and only if Λ is lacunary. We shall discuss these cases in Sect. 9.2

Here we turn now to the quasilacunary case.

Proposition 6.3.3 *Let Λ be quasilacunary such that there are indices $0 = n_0 < n_1 < \ldots$, a positive integer N and $q > 1$ with $n_{m+1} - n_m \le N$, $m = 0, 1, \ldots$, and $\inf_{m \ge 1} \lambda_{n_m+1}/\lambda_{n_m} \ge q$. Then there is a universal constant $c > 0$ such that, with $f = \sum_k \alpha_k t^{\lambda_k} \in [M(\Lambda)]_E$ and $f_m = \sum_{k=n_m+1}^{n_{m+1}} \alpha_k t^{\lambda_k}$, we have*

$$\|f_m\|_E \le c\|f\|_E \quad \text{for all } m .$$

Actually,

$$c = 2^{N^2+N}(\sup\{4^M \|T_{2^{-1/\lambda_n}}\|^M (1 + \|T_{2^{-1/\lambda_n}}\|)^M : n = 1, 2, \ldots\})^{N^2}$$

where M is a positive number satisfying

$$N\sum_{k=1}^\infty \left(\eta\left(\frac{1}{2^{1/q^k}}\right)^M + \eta\left(\frac{1}{2^{q^k}}\right)^M\right) \le \frac{1}{2} .$$

Proof. We assume without loss of generality that $n_{l+1} - n_l = N$ for all l. (If this is not the case then increase Λ suitably.)

Fix m. Consider k with $n_m < k \leq n_{m+1}$. Let $j \leq n_m$ or $j \geq n_{m+1} + 1$ and consider l such that $n_l + 1 \leq j \leq n_{l+1}$. Since Λ is quasilacunary we obtain

$$\frac{\lambda_j}{\lambda_k} \leq q^{l-m} \text{ if } l < m \quad \text{and} \quad \frac{\lambda_j}{\lambda_k} \geq q^{l-m} \text{ if } l > m \, .$$

As before we find $M > 0$ independent of f and m such that

$$\sum_{j \leq n_m} \eta(2^{-\lambda_j/\lambda_k})^M + \sum_{j \geq n_{m+1}+1} \eta(2^{-\lambda_j/\lambda_k})^M$$

$$\leq N \sum_{l=1}^{m-1} \eta(2^{-q^{l-m}})^M + N \sum_{l=m+1}^{\infty} \eta(2^{-q^{l-m}})^M \leq \frac{1}{2}$$

since $\eta(y)$ is increasing if $0 < y < 1/2$ and decreasing if $1/2 < y < 1$. Put

$$b = \sup\{4^M \|T_{2^{-1/\lambda_n}}\|^M (1 + \|T_{2^{-1/\lambda_n}}\|)^M : n = 1, 2, \ldots\} \, .$$

a) At first we show the following:

Let $n_m + 1 \leq k \leq n_{m+1}$ and

$$\tilde{\Lambda} = \{\lambda_1, \ldots, \lambda_{n_m}, \lambda_k, \lambda_{n_{m+1}+1}, \lambda_{n_{m+1}+2}, \ldots\}$$

Then, for any $f = \sum_{\lambda_j \in \tilde{\Lambda}} \alpha_j t^{\lambda_j} \in [M(\tilde{\Lambda})]_E$ we have

$$\|\alpha_k t^{\lambda_k}\|_E \leq 2^{N-1} b^N \|f\|_E$$

Indeed, put

$$\Lambda_j = \{ \lambda_{n_l+j} : l = 1, 2, \ldots, m-1, m+1, m+2, \ldots\} \cup \{\lambda_k\}$$

where $j = 1, \ldots, N$. Moreover, put

$$W_j = \overline{\text{span}}\{t^{\lambda_l} : l = n_h+j+1, \ldots, n_{h+1}, \ h = 1, \ldots, m-1, m+1, m+2, \ldots\} \, ,$$

$j = 1, \ldots, N-1$, and $W_N = \{0\}$.

Λ_1 is lacunary. Moreover we have $\|T_\rho\|_{E/W_1} \leq \|T_\rho\|_E$ for any $\rho \in [0, 1]$. Hence 6.3.2 yields

$$\|\alpha_k t^{\lambda_k} + W_1\|_{E/W_1} \leq b\|f + W_1\|_{E/W_1} \leq b\|f\|_E$$

Find $w_1 = g + w_2$ with $g = \sum_{l \neq m} \beta_l t^{\lambda_{n_l+2}}$ and $w_2 \in W_2$ satisfying

$$\|\alpha_k t^{\lambda_k} + w_1\|_E \leq 2\|\alpha_k t^{\lambda_k} + W_1\|_{E/W_1}$$

and therefore

$$||\alpha_k t^{\lambda_k} + g + W_2||_{E/W_2} \leq ||\alpha_k t^{\lambda_k} + w_1||_E \leq 2b||f||_E$$

Λ_2 is lacunary. Hence we obtain, by 6.3.2,

$$||\alpha_k t^{\lambda_k} + W_2||_{E/W_2} \leq 2b||\alpha_k t^{\lambda_k} + g + W_2||_{E/W_2} \leq 2b^2||f||_E$$

Similarly we infer

$$||\alpha_k t^{\lambda_k} + W_3||_{E/W_3} \leq 2^2 b^3 ||f||_E$$

Continuation yields after N steps

$$||\alpha_k t^{\lambda_k}||_E \leq 2^{N-1} b^N ||f||_E \leq 3^N 8^{MN} b^N ||f||_E$$

b) Let f and f_m be as in the statement of the proposition and assume $\alpha_{j+1} = \ldots = \alpha_{n_{m+1}} = 0$ for some $j \geq n_m + 1$. Then we prove the proposition by induction on j. If $j = 1$ then a) yields

$$||f_m||_E = ||\alpha_{n_m+1} t^{\lambda_{n_m+1}}||_E \leq 2^{N-1} b^N ||f||_E$$

which proves the claim. Put $c_1 = 2^{N-1} b^N$.

Now assume we have $||f_m||_E \leq c_{j-1}||f||_E$ if $f_m = \sum_{n_m+1}^{j-1} \alpha_k t^{\lambda_k}$ where c_{j-1} is independent of m and f.

Then go over to the case $f_m = \sum_{n_m+1}^{j} \alpha_k t^{\lambda_k}$ and put

$$V = \text{span}\{t^{\lambda_{n_m+1}}, \ldots, t^{\lambda_{j-1}}\} .$$

Moreover, let

$$\tilde{\Lambda}(j) = \{\lambda_1, \ldots, \lambda_{n_m}, \lambda_j, \lambda_{n_{m+1}+1}, \lambda_{n_{m+1}+2}, \ldots\}$$

and

$$\Lambda(j) = \tilde{\Lambda}(j) \cup \{\lambda_{n_m+1}, \ldots, \lambda_{j-1}\}$$

Put $d = ||t^{\lambda_j} + V||_{E/V}^{-1}$ and fix $v \in V$ with $||dt^{\lambda_j} + v||_E \leq 2$. According to a) applied to the norm in E/V we find $x^* \in ([M(\Lambda(j))]_{E/V})^*$ with $x^*(dt^{\lambda_j} + V) = 1$ and $x^*(t^{\lambda_k} + V) = 0$ whenever $k \neq j$, $\lambda_k \in \tilde{\Lambda}(j)$, where $||x^*|| \leq 2^{N-1} b^N$. x^* can be interpreted as element of $[M(\Lambda(j))]_E^*$ vanishing on all t^{λ_k}, $\lambda_k \in \Lambda(j) \setminus \{\lambda_j\}$. Now we find β_k such that

$$f_m(t) = \sum_{k=n_m+1}^{j} \alpha_k t^{\lambda_k} = \sum_{k=n_m+1}^{j-1} \beta_k t^{\lambda_k} + \beta_j (dt^{\lambda_j} + v)$$

and $\beta_j = x^*(f)$. Put $h_m = \sum_{k=n_m+1}^{j-1} \beta_k t^{\lambda_k}$. Then, by the induction hypothesis, we obtain

$$\begin{aligned}
||f_m||_E &\leq ||h_m||_E + 2|x^*(f)| \\
&\leq c_{j-1}||f - x^*(f)(dt^{\lambda_j} + v)||_E + 2|x_j^*(f)| \\
&\leq c_{j-1}||f||_E + (2c_{j-1} + 2)|x_j^*(f)| \\
&\leq (c_{j-1} + 2^{N-1} b^N (2c_{j-1} + 2))||f||_E .
\end{aligned}$$

We finish the induction by taking

$$c_j = 2^{N+1}b^N c_{j-1} \geq c_{j-1} + 2^{N-1}b^N(2c_{j-1}+2) .$$

(Recall $b > 1$.) Then we have

$$c_j \leq (2^{N+1}b^N)^j .$$

Hence $c_N = 2^{N^2+N}b^{N^2}$.   □

We finish this section with a corollary which we will apply in 9.3.

Corollary 6.3.4 *Consider* $t_0 = 0 < t_1 < t_2 < \ldots$ *with* $\lim_{k\to\infty} t_k = 1$. *Under the hypothesis of 6.3.3, if* $E = L_p$ *for* $1 \leq p < \infty$, *there is a constant* $c > 0$ *such that*

$$\left(\sum_{k=1}^{\infty} \int_{t_{k-1}}^{t_k} |f_k|^p dt\right)^{1/p} \leq c\|f\|_{L_p} \quad \text{for all } f \in L_p .$$

Proof. Fix m and let $E = L_p$ be endowed with the seminorm

$$\|f\|_m = \left(\sum_{j=0}^{\infty} \frac{1}{2^j} \sup_{2^{-j-1} \leq \tau \leq 2^{-j}} \int_{\tau t_{m-1}}^{\tau t_m} |f|^p dt\right)^{1/p}$$

$(E, \|\cdot\|_m)$ clearly satisfies our general assumptions. If $\rho \in \,]0,1[$ is such that $2^{-k-1} \leq \rho < 2^{-k}$ for some k then we have

$$\|T_\rho f\|_m = \left(\frac{1}{\rho} \sum_{j=0}^{\infty} \frac{1}{2^j} \sup_{2^{-j-1} \leq \tau \leq 2^{-j}} \int_{\tau \rho t_{m-1}}^{\tau \rho t_m} |f|^p dt\right)^{1/p}$$

$$\leq \left(\frac{2^{k+1}}{\rho} \sum_{j=0}^{\infty} \frac{1}{2^j} \sup_{2^{-j-1} \leq \tau \leq 2^{-j}} \int_{\tau t_{m-1}}^{\tau t_m} |f|^p dt\right)^{1/p}$$

$$\leq \left(\frac{2}{\rho^2}\right)^{1/p} \|f\|_m$$

Hence $\|T_\rho\|_m \leq (2/\rho^2)^{1/p}$. 6.3.3 yields a constant $c > 0$ with

$$\left(\int_{t_{m-1}}^{t_m} |f_m|^p dt\right)^{1/p} \leq \|f_m\|_m \leq c\|f\|_m$$

for all m. Let $\tau_j \in [2^{-j-1}, 2^{-j}]$ be such that

$$\int_{\tau_j t_{m-1}}^{\tau_j t_m} |f|^p dt = \sup_{2^{-j-1} \leq \tau \leq 2^{-j}} \int_{\tau t_{m-1}}^{\tau t_m} |f|^p dt .$$

Then we obtain

$$\left(\sum_{m=1}^{\infty}\int_{t_{m-1}}^{t_m}|f_m|^p dt\right)^{1/p} \le c\left(\sum_{m=1}^{\infty}||f||_m^p\right)^{1/p}$$

$$= c\left(\sum_{j=0}^{\infty}\frac{1}{2^j}\sum_{m=1}^{\infty}\int_{\tau_j t_{m-1}}^{\tau_j t_m}|f|^p dt\right)^{1/p}$$

$$\le c\left(\sum_{j=0}^{\infty}\frac{1}{2^j}\int_0^{\tau_j}|f|^p dt\right)^{1/p}$$

$$\le 2c||f||_{L_p}$$

$\square$

6.4 On $\bar{c}$-Completeness of Müntz Sequences

Here we restrict ourselves to the sup-norm $||\cdot||_C$ on $[0,1]$ and consider $E = C_0[0,1]$. In 6.2 we found coefficient estimates for Müntz polynomials $f(t) = \sum_k \alpha_k t^{\lambda_k}$ of the form $|\alpha_k| \le a^{\lambda_k}||f||_C$ for suitable $a > 1$ provided that $\sum_{k=1}^{\infty} 1/\lambda_k < \infty$. Now we consider the initial Müntz sequence $M(\Lambda)$ where

$$\Lambda = \{1, 2, 3, \ldots\}\,.$$

In view of 6.1.5 $M(\Lambda)$ is not a minimal system and hence we do not have similar estimates for this Λ. It turns out that in connection with uniform polynomial approximation of a function $f \in C_0$ even the approximating polynomials cannot always be arranged to have such "good" coefficient estimates.

In the following let $\bar{c} = \{c_k\}_{k=1}^{\infty}$ be a sequence of positive numbers with $c_k \uparrow \infty$.

We say that $f \in C_0$ admits a $\bar{c}$-approximation if, for each $\epsilon > 0$, there is a polynomial $P_\epsilon(t) = \sum_k \beta_k t^k$ with $||f - P_\epsilon||_C \le \epsilon$ and $|\beta_k| \le c_k ||f||_C$ for all k. (Compare this with the definition of $\bar{c}$-completeness preceding 2.1.7.)

Proposition 6.4.1 *[56] Let $f \in C_0$ be such that there is α, $0 < \alpha < 1$, with $f(t) = 0$ for $t \in [0, \alpha]$. Then there is some $a = a(\alpha) > 0$ such that f admits a $\bar{c}$-approximation whenever*

$$\liminf_{k\to\infty}\frac{a^k}{c_k} = 0\,.$$

Proof. We claim that $a = 6^{1/\alpha}$ satisfies the assertion of the proposition (although this value of a is not the best possible).

Indeed, consider $c_k > 0$ with $\liminf_k a^k/c_k = 0$. Then we find a subsequence $\{k_s\}_{s=1}^{\infty}$ with $\lim_{s\to\infty} a^{k_s}/c_{k_s} = 0$. We consider the positive integers n_s with $k_s = [\alpha n_s] + 1$, $s = 1, 2, \ldots$ ($[x]$ is the largest integer $\le x$ for $x \in \mathbf{R}$).

At first we estimate the coefficients of the Bernstein polynomials for the function $f(t)$:

$$B_{n_s}(t) = \sum_{k=0}^{n_s} \binom{n_s}{k} f\left(\frac{k}{n_s}\right) t^k (1-t)^{n_s-k} = a_o + a_1 t + \ldots + a_{n_s} t^{n_s} ,$$

where, by virtue of the fact that $f(k/n_s) = 0$, $k = 0, 1, \ldots, k_s - 1$, the coefficients of the monomials different from zero begin with that for t^{k_s}, i.e.

$$B_{n_s}(t) = a_{k_s} t^{k_s} + \ldots + a_{n_s} t^{n_s} .$$

Put $A = \max_{0 \leq t \leq 1} |f(t)|$. The coefficients of $B_{n_s}(t)$ do not exceed, in absolute values, those of the polynomial

$$P_{n_s}(t) = A \sum_{k=k_s}^{n_s} \binom{n_s}{k} t^k (1+t)^{n_s-k} ,$$

and the coefficients of $P_{n_s}(t)$, in turn, do not exceed those of the polynomial

$$\tilde{P}_{n_s}(t) = A \cdot 2^{n_s} \sum_{k=k_s}^{n_s} t^k (1+t)^{n_s-k} .$$

It is not hard to see that an arbitrary coefficient of $\tilde{P}_{n_s}$ does not exceed $A \cdot 2^{n_s} \cdot 2^{n_s} < A \cdot 6^{n_s}$. Hence,

$$|a_k| < A \cdot 6^{n_s}, \quad k = k_s, k_s + 1, \ldots, n_s .$$

Since $k_s = [\alpha n_s] + 1$ we obtain $k_s > \alpha n_s$ and $n_s < k_s/\alpha$. So $|a_k| < A6^{k_s/\alpha}$, $k = k_s, \ldots, n_s$. Thus, with $a = 6^{1/\alpha}$, $|a_k| < Aa^{k_s}$, $k = k_s, \ldots, n_s$. Since $\lim_{s \to \infty} a^{k_s}/c_{k_s} = 0$ we find an s_0 with $Aa^{k_s} < c_{k_s}$ for all $s > s_0$ and, consequently, $|a_k| < c_{k_s}$, $k = k_s, \ldots, n_s$. But then, in view of the monotonicity of $\bar{c}$,

$$|a_k| < c_k, \quad k = k_s, k_s + 1, \ldots, n_s .$$

Moreover, from the approximation property of Bernstein polynomials [112], we obtain

$$\lim_{s \to \infty} \|f - P_{n_s}\|_C = 0 .$$

$\square$

Lemma 6.4.2 *If the function $f \in C_0$ admits a $\bar{c}$-approximation where $c_k < Aa^k$, $k = 1, 2, \ldots$, for some $A > 0$ and $a \geq 1$, then $f(t)$ is analytically extendable to the disk $|z| < 1/a$.*

Proof. For arbitrary $\epsilon > 0$ let

$$P_{n(\epsilon)}(t) = a_1^{(\epsilon)} t + \ldots + a_{n(\epsilon)}^{(\epsilon)} t^{n(\epsilon)}$$

be a polynomial for which

$$\max_{0\le t\le 1} |f(t) - P_{n(\epsilon)}(t)| < \epsilon, \quad |a_k^{(\epsilon)}| \le c_k \le Aa^k, \; k = 1, 2, \ldots, n(\epsilon) \, .$$

For a given $\delta > 0$ we now show that all the $P_{n(\epsilon)}(z)$ are uniformly bounded in the disk $|z| < (1/a) - \delta$. We have

$$
\begin{aligned}
|P_{n(\epsilon)}(z)| &\le Aa\left(\frac{1}{a} - \delta\right) + Aa^2\left(\frac{1}{a} - \delta\right)^2 + \ldots + Aa^n\left(\frac{1}{a} - \delta\right)^n \\
&\le \frac{Aa(\frac{1}{a} - \delta)}{1 - a(\frac{1}{a} - \delta)} \\
&= \frac{A(1 - a\delta)}{a\delta} \, .
\end{aligned}
$$

By the theorem of Montel [112] it is possible to select a subsequence $\{P_{n(\epsilon_k)}(z)\}_{k=1}^{\infty}$ converging uniformly to an analytic function $h(z)$ in the disk $|z| < (1/a) - 2\delta$ for $k \to \infty$, where $\epsilon_k \to 0$. In addition we obtain $h(t) = f(t)$ if $t \in [0, (1/a) - 2\delta]$. Since $\delta > 0$ can be taken arbitrarily small we obtain a function $h(z)$, analytic on the disk $|z| < 1/a$, such that $h(t) = f(t)$ for all $t \in [0, 1/a[$. □

Since there are functions in C_0 which cannot be extended to an analytic function on a disk, 6.4.1 and 6.4.2 imply

Theorem 6.4.3 $\{t^k\}_{k=1}^{\infty}$ *is $\bar{c}$-complete in $C_0[0,1]$ if and only if, for any $a >$ 1, the condition*

$$\liminf_{k\to\infty} \frac{a^k}{c_k} = 0$$

is satisfied.

Theorems 6.4.3 and 2.7.3 imply the stability of the initial degree sequence in C:

Theorem 6.4.4 *If the sequence $\bar{\epsilon} = \{\epsilon_k\}_{k=1}^{\infty}$ satisfies the condition*

$$\lim_{k\to\infty} a^k \epsilon_k = 0 \quad \text{for each } a > 1$$

then the completeness of $\{t^k\}_{k=1}^{\infty}$ is $\bar{\epsilon}$-stable, i.e. each $\bar{g} = \{g_k\}_{k=1}^{\infty} \subset C_0[0,1]$, with $\|g_k - t^k\|_C \le \epsilon_k$ for all k, is complete in C_0, too.

Proof. By assumption, for any $a > 1$ we have $\lim_{n\to\infty} 3^k a^k \epsilon_k = 0$. Put $c_k = 3^{-k}\epsilon_k^{-1}$. Then $\{t^k\}_{k=1}^{\infty}$ is $\bar{c}$-complete and we obtain $\sum_{k=1}^{\infty} \epsilon_k c_k < 1$. Now 2.7.3 completes the proof. □

6.5 Coefficient Estimates on $[a, b]$ where $a > 0$

In comparison with the coefficient estimates for polynomial approximation on $[0, 1]$ we obtain considerable improvements if we restrict ourselves to closed intervals which do not contain 0.

Theorem 6.5.1 *Let $0 < a < b < 1$ and $\lambda > 0$. Then, for every $\epsilon > 0$ and every $f \in C[a, b]$, there is a function g of the form*

$$g(t) = \sum_{k=1}^{n} \alpha_k t^{-\lambda k}, \ t \in [a, b], \quad \text{for some } \alpha_k \text{ and } n \ ,$$

satisfying

$$\sup_{a \leq t \leq b} |f(t) - g(t)| < \epsilon \quad \text{and} \quad |\alpha_k| \leq \epsilon \ \text{ for all } k \ .$$

Proof. Let $\mathbf{T} = \{z \in \mathbf{C} : |z| = 1\}$ and consider $K := [a, b] \cup \mathbf{T}$ as a subset of $\mathbf{C}$. Define $F_0 : K \to \mathbf{R}$ by

$$F_0(t) = \tfrac{1}{2}(t^\lambda + t^{-\lambda}) \ \text{ if } \ t \in [a, b] \ ,$$

$$F_0(z) = \tfrac{1}{2}(z + \bar{z}) \qquad \text{if } z \in \mathbf{T} \ .$$

Of course, F_0 is continuous on K. Consider the equivalence relation $\sim$ on K defined by

$$k_1 \sim k_2 \text{ if and only if } F_0(k_1) = F_0(k_2) \ .$$

Put $X = \text{closed span } \{F_0^k \ : \ k = 0, 1, 2, \ldots\}$. By the Weierstraß-Stone theorem we obtain that X is equal to $C(K/\sim)$, the space of all continous functions on K which identify k_1 and k_2 in K whenever $k_1 \sim k_2$. Now, let $z, w \in \mathbf{T}$. Then we have

$$z \sim w \text{ if and only if } \ \text{Re } z = \text{Re } w \text{ if and only if } z = w \text{ or } z = \bar{w} \ .$$

Moreover, for $s, t \in [a, b]$, we obtain $s \sim t$ if and only if $s = t$. This follows from the fact that the function $x \mapsto x + x^{-1}$ is strictly decreasing on $[a, b]$ (recall, $b < 1$). Finally, if $t \in [a, b]$, $z \in \mathbf{T}$, then $F_0(t) \neq F_0(z)$. Indeed, otherwise we would have $t^\lambda + t^{-\lambda} = z + \bar{z}$ and hence

$$t^\lambda = \frac{z + \bar{z}}{2} \pm \sqrt{\left(\frac{z + \bar{z}}{2}\right)^2 - 1} \ .$$

This is impossible since $0 < t^\lambda < 1$ but the right-hand side is either ± 1 or non-real.

Now, fix $\epsilon > 0$ and $f \in C[a, b]$. Put $\delta = \min((1 - b^\lambda)(3b^\lambda)^{-1}\epsilon, 3^{-1}\epsilon)$. Define $F \in C(K/\sim)$ by

$$F(t) = f(t) \text{ if } t \in [a, b] \,,$$

$$F(z) = 0 \quad \text{if } z \in \mathbf{T} \,.$$

Find a function G of the form $G = \sum_{k=0}^{n} \gamma_k F_0^k$ for some γ_k with $\sup_{k \in K} |F(k) - G(k)| < \delta$. In particular

$$G(t) = \sum_{k=0}^{n} \gamma_k \left(\frac{t^\lambda + t^{-\lambda}}{2}\right)^k \text{ if } t \in [a, b] \,,$$

$$G(z) = \sum_{k=0}^{n} \gamma_k \left(\frac{z + \bar{z}}{2}\right)^k \quad \text{if } z \in \mathbf{T} \,.$$

Hence, for suitable α_k, β_k,

$$G(t) = \alpha_0 + \sum_{k=1}^{n} (\alpha_k t^{-\lambda k} + \beta_k t^{\lambda k}) \text{ if } t \in [a, b] \,,$$

$$G(z) = \alpha_0 + \sum_{k=1}^{n} (\alpha_k \bar{z}^k + \beta_k z^k) \quad \text{if } z \in \mathbf{T} \,.$$

Since $\sup_{z \in \mathbf{T}} |G(z)| < \delta$ we obtain $|\alpha_k| < \delta$, $k = 0, 1, \ldots, n$, and $|\beta_k| < \delta$, $k = 1, 2, \ldots, n$. In view of $\sup_{t \in [a,b]} |F(t) - G(t)| < \delta$ we conclude

$$\sup_{t \in [a,b]} \left| f(t) - \sum_{k=1}^{n} (\alpha_k t^{-\lambda k} + \beta_k t^{\lambda k}) \right| < 2\delta \leq \frac{2}{3}\epsilon \,.$$

Finally,

$$\left| \sum_{k=1}^{n} \beta_k t^{\lambda k} \right| \leq \frac{b^\lambda \delta}{1 - b^\lambda} \leq \frac{\epsilon}{3} \text{ since } t^\lambda \leq b^\lambda \,.$$

So,

$$\sup_{t \in [a,b]} \left| f(t) - \sum_{k=1}^{n} \alpha_k t^{-\lambda k} \right| < \epsilon \,.$$

$\square$

Corollary 6.5.2 *(a) Let $1 < c < d$, $\lambda > 0$ and $f \in C[c, d]$. Then, for every $\epsilon > 0$, we find a function of the form*

$$g(t) = \sum_{k=1}^{n} \alpha_k t^{\lambda k}, \ t \in [c, d] \,,$$

with

$$\sup_{t \in [c,d]} |f(t) - g(t)| < \epsilon \text{ and } |\alpha_k| < \epsilon, \ k = 1, 2, \ldots, n \,.$$

(b) Let $0 < a < b$, $\lambda > 0$ and $f \in C[a, b]$. Then, for every $\epsilon > 0$, we find a function of the form

$$g(t) = \sum_{k=1}^{n} \alpha_k t^{\lambda k}, \ t \in [a, b] \,,$$

with

$$\sup_{t\in[a,b]} |f(t) - g(t)| < \epsilon \quad and \quad |\alpha_k| < \frac{\epsilon}{((1-\epsilon)a)^{\lambda k}}, \quad k = 1, 2, \ldots, n \, .$$

Proof. (a) follows from 6.5.1 by substituting $t = s^{-1}$ (with $a = d^{-1}$ and $b = c^{-1}$).

(b) follows from (a) by substituting $t = a(1 - \epsilon)s$ (with $c = (1 - \epsilon)^{-1}$, $d = ba^{-1}(1 - \epsilon)^{-1}$). □

Corollary 6.5.3 *Let X be a Banach space with a normalized complete system $\{e_k\}_{k=1}^\infty$. Furthermore, assume that $0 < a < b$ and $\lambda > 0$.*
(a) Then, for any $\epsilon > 0$,

$$\{(t^{\lambda k}, a^{\lambda k}(1 - \epsilon)^{\lambda k} e_k) \ : \ k = 1, 2, \ldots\}$$

is a complete system for $C[a, b] \oplus X$.
(b) For any $\epsilon > 0$ and any $p \in [1, \infty[$,

$$\{(t^{\lambda k}, a^{\lambda k}(1 - \epsilon)^{\lambda k} e_k) \ : \ k = 1, 2, \ldots\}$$

is a complete system for $L_p[a, b] \oplus X$.

Remark. Note that

$$\{(t^{\lambda k}, a^{\lambda k}(1 - \epsilon)^{\lambda k} e_k) \ : \ k = 1, 2, \ldots\}$$

is always minimal provided that $\{e_k\}_{k=1}^\infty$ is minimal in X. If $a > 1$, then one can show with similar arguments that even

$$\{(t^{\lambda k}, \epsilon e_k) \ : \ k = 1, 2, \ldots\}$$

is a complete system for $C[a, b] \oplus X$ and $L_p[a, b] \oplus X$, resp. This system is uniformly minimal if $\{e_k\}_{k=1}^\infty$ is uniformly minimal in X.

Proof of the Corollary. Since $C[a, b]$ is dense in $L_p[a, b]$ we only have to show that, for any $f \in C[a, b]$,

$$(f, 0) \in \text{ closed span}\{(t^{\lambda k}, a^{\lambda k}(1 - \epsilon)^{\lambda k} e_k) \ : \ k = 1, 2, \ldots\}$$

This follows from the fact that, in view of 6.5.2 (b), for any $\delta > 0$, we find

$$G(t) = \sum_{k=1}^n \alpha_k(t^{\lambda k}, a^{\lambda k}(1 - \epsilon)^{\lambda k} e_k)$$

with

$$\left| f(t) - \sum_{k=1}^n \alpha_k t^{\lambda k} \right| < \delta, \ t \in [a, b], \quad and \quad |\alpha_k| < \frac{\delta}{(1 - \delta)^{\lambda k} a^{\lambda k}} \, .$$

So

$$\left\| \sum_{k=1}^n \alpha_k a^{\lambda k}(1 - \epsilon)^{\lambda k} e_k \right\| < \delta \frac{(\frac{1-\epsilon}{1-\delta})^\lambda}{1 - (\frac{1-\epsilon}{1-\delta})^\lambda} = \delta \frac{(1 - \epsilon)^\lambda}{(1 - \delta)^\lambda - (1 - \epsilon)^\lambda} \, .$$

This concludes the proof since $\delta < \epsilon$ can be chosen arbitrarily small. □

7

Classification and Elementary Properties of Müntz Sequences

Again we consider $\Lambda = \{\lambda_k\}_{k=1}^\infty$ satisfying $0 < \lambda_1 < \lambda_2 < \ldots$ and $\sum_{n=1}^\infty 1/\lambda_k < \infty$. At first we will be concerned with different classes of Λ which are distinguished by special properties. Then we study the underlying Müntz polynomials $\sum_{k=1}^n \alpha_k t^{\lambda_k}$. In particular we give estimates for the inclination of the elements t^{λ_k} as well as of the differences $t^{\lambda_k} - t^{\lambda_{k+1}}$. One of the main results of this chapter is Theorem 7.4.4 where we show that Λ is non-lacunary if and only if $M(\Lambda)$ is closing. For E we always consider C_0 or L_p.

7.1 Different Classes of Λ

We start with

Definition 7.1.1 *A sequence Λ satisfying $0 < \lambda_1 < \lambda_2 < \ldots$ (as well as the corresponding sequence $M(\Lambda)$ and the Müntz space $[M(\Lambda)]_E$) will be called*
1. standard , *if*
$$\lim_{k \to \infty} \frac{\lambda_{k+1}}{\lambda_k} = 1 \,,$$
2. rational , *if all λ_k are non-negative rationals,*
3. integer, *if all λ_k are non-negative integers,*
4. sparse *or* non-dense , *if $\sum_k 1/\lambda_k < \infty$,*
5. dense , *if $\sum_k 1/\lambda_k = \infty$.*

In 6.3.1 we already introduced lacunary and quasilacunary sequences. Now we extend this notion.

Definition 7.1.2 *Λ will be called* block lacunary *if, for some increasing sequence of integers $\bar{n} = \{n_k\}_{k=1}^\infty$ and some $\beta > 1$, we have $\lambda_{n_k+1}/\lambda_{n_k} \geq \beta$, $k = 1, 2, \ldots$*

If Λ is block lacunary with respect to $\bar{n}$ and β we also speak of a $(\bar{n}, \beta)$-block lacunary sequence. The intervals

$$I_k = \{m : m \ an\ integer\ ,\ n_k + 1 \leq m \leq n_{k+1}\}$$

will be called block intervals.

Recall that, according to 6.3.1, Λ is quasilacunary if it is $(\bar{n}, \beta)$-block lacunary for some $\beta > 1$ and we have $\sup_k(n_{k+1} - n_k) < \infty$. Moreover, Λ is lacunary if it is $(\bar{n}, \beta)$-block lacunary with $n_{k+1} - n_k = 1$ for all k.

Proposition 7.1.3 *The following are equivalent*
(i) Λ is quasilacunary
(ii) There are lacunary $\Lambda_1, \ldots, \Lambda_m$ such that $\Lambda = \cup_{i=1}^m \Lambda_i$
(iii) For arbitrary $\beta > 1$ there is N such that $\Lambda \cap [\beta^j, \beta^{j+1}]$, $j = 1, 2, \ldots$, has at most N elements
(iv) There is an increasing sequence of integers $\bar{n} = \{n_k\}_{k=1}^\infty$ and some $\beta > 1$ such that

$$\lambda_{n_{k+1}}/\lambda_{n_k} \geq \beta, \ k = 1, 2, \ldots, \quad and \quad \sup_k(n_{k+1} - n_k) < \infty \,.$$

Proof. $(i) \Rightarrow (ii)$: If we fix exactly one λ_j in each block we obtain finitely many lacunary Λ_i satisfying (ii).
$(ii) \Rightarrow (iii)$: Fix an arbitrary $\beta > 1$. Since all Λ_j are lacunary there are $N_j > 0$ such that $\Lambda_j \cap [\beta^k, \beta^{k+1}]$ has at most N_j elements for all k. Put $N = \sup_{j=1,\ldots,m} N_j$.
$(iii) \Rightarrow (iv)$: Fix $\beta > 1$ and put $m_k = \sup\{i : \lambda_i \leq \beta^k\}$. Then, by (iii), $\sup_k(m_{k+1} - m_k) < \infty$. We may assume $\Lambda \cap]\beta^k, \beta^{k+1}] \neq \emptyset$ for each k, otherwise enlarge Λ. Hence we have $m_k \neq m_{k+1}$ and $\beta^k \leq \lambda_{m_k+1}, \ldots, \lambda_{m_{k+1}} \leq \beta^{k+1}$. Put $n_k = m_{2k}$. Then $\sup_k(n_{k+1} - n_k) < \infty$ and

$$\frac{\lambda_{n_{k+1}}}{\lambda_{n_k}} \geq \frac{\beta^{2k+1}}{\beta^{2k}} = \beta \,.$$

$(iv) \Rightarrow (i)$: By assumption we have

$$\prod_{j=n_k}^{n_{k+1}-1} \left(\frac{\lambda_{j+1}}{\lambda_j}\right) = \frac{\lambda_{n_{k+1}}}{\lambda_{n_k}} \geq \beta$$

and the numbers of the factors in the preceding products are uniformly bounded. Therefore we find $\delta > 1$ and, for each k, some index m_k such that $n_k \leq m_k \leq n_{k+1}$ and $\lambda_{m_k+1}/\lambda_{m_k} \geq \delta$. Since $\sup_k(n_{k+1} - n_k) < \infty$ we also have $\sup_k(m_{k+1} - m_k) < \infty$. Now, Λ is quasilacunary with respect to $\bar{m} = \{m_k\}_{k=1}^\infty$ and δ.   □

We also note that block lacunary and standard are opposite properties.

Proposition 7.1.4 *The following are equivalent*
(i) Λ is not block lacunary
(ii) Λ is standard

Proof. $(i) \Rightarrow (ii)$: Otherwise find $\delta > 0$ and indices $n_1 < n_2 < \ldots$ with $\lambda_{n_k+1}/\lambda_{n_k} \geq 1 + \delta$ for all k.
$(ii) \Rightarrow (i)$ is obvious. $\qquad\qquad\qquad\qquad\qquad\qquad\qquad\qquad\qquad\qquad\qquad$ □

We conclude this section with two more classes of Λ.

Definition 7.1.5 *1. Let, for some $a > 0a$, $\delta > 0$ and $s \geq 1$*

$$\lambda_k = a + \delta k^s, \qquad k = 1, 2, \ldots$$

Then we call Λ an s-arithmetic sequence. If $s = 1$ then Λ is simply called an arithmetic *sequence.*
2. Let $\bar{n} = \{n_k\}_{k=1}^{\infty}$ be an increasing sequence of indices and assume that there are numbers $a_k > 0$ and $\delta_k > 0$ with

$$\lambda_j = a_k + (j - n_k)\delta_k, \qquad n_k + 1 \leq j \leq n_{k+1}, \qquad k = 1, 2, \ldots$$

Put $\bar{a} = \{a_k\}_{k=1}^{\infty}$ and $\bar{\delta} = \{\delta_k\}_{k=1}^{\infty}$. Then Λ will be called a $(\bar{n}, \bar{a}, \bar{\delta})$-block arithmetic sequence.

Of course, an s-arithmetic sequence is non-dense if and only if $s > 1$.

Virtually nothing is known about the Banach space $[M(\{k^s\}_{k=1}^{\infty})]_C$ if $s > 1$. On the other hand, in 9.3 we will give a complete Banach space characterization of $[M(\Lambda)]_E$ for $E = C$ and $E = L_p$, $1 \leq p < \infty$, if Λ is quasilacunary. (Then $[M(\Lambda)]_C \sim c_0$ and $[M(\Lambda)]_{L_p} \sim l_p$.) Moreover, in 10.2 we show that there is a block lacunary Λ where $[M(\Lambda)]_C$ is not isomorphic to c_0.

Definition 7.1.6 *1. Let, for some $a > 0$ and $q > 0$, $\lambda_k = aq^k$, $k = 1, 2, \ldots$ Then we call Λ a* geometric *(or (a, q)-geometric) sequence.*
2. Let $\bar{n} = \{n_k\}_{k=1}^{\infty}$ be an increasing sequence of indices and assume that there are numbers $a_k > 0$ and $q_k > 0$ with

$$\lambda_j = a_k q_k^{j - n_k}, \qquad n_k + 1 \leq j \leq n_{k+1}, \qquad k = 1, 2, \ldots$$

Put $\bar{a} = \{a_k\}_{k=1}^{\infty}$ and $\bar{q} = \{q_k\}_{k=1}^{\infty}$. Then Λ will be called a $(\bar{n}, \bar{a}, \bar{q})$-block geometric sequence.

7.2 Iterated Differences

For the sequence Λ let $d\Lambda$ denote the differences $d\Lambda = \{\lambda_{k+1} - \lambda_k\}_{k=1}^{\infty}$. Then go on to define in the same fashion $d^2(\Lambda) = d(d\Lambda)$, $d^3(\Lambda) = d(d^2(\Lambda))$ etc. Put $d^0(\Lambda) = \Lambda$.

Definition 7.2.1 *The sequence Λ will be called*
1. k-regular *(or strictly* k-regular*) if, for some positive integer k, all sequences $d^j(\Lambda)$, $j = 0, 1, \ldots, k$, consist of non-negative (or strictly positive) numbers,*
2. absolutely monotone *(or strictly absolutely monotone), if Λ is k-regular (or strictly k-regular) for all positive integers k.*

For example, the sequence $\Lambda = \{q^k\}_{k=1}^{\infty}$ is strictly absolutely monotone for any $q > 1$. It will turn out that strictly absolutely monotone sequences are always block lacunary but not necessarily lacunary.

Proposition 7.2.2 *Let Λ be a strictly absolutely monotone sequence of integers. Then we have $\lambda_k \geq 2^{k-1}$ for all k. Moreover, Λ is block-lacunary.*

Proof. Let $\Lambda = \{\lambda_k\}_{k=1}^{\infty}$ be a strictly absolutely monotone sequence of integers. Define $a_{1,k} = \lambda_k$, $k = 1, 2, \ldots$ and, by induction, $a_{m,k} = a_{m-1,k} - a_{m-1,k-1}$, $k = m, m+1, \ldots$. Then all $a_{m,k}$ are positive integers. In particular, $a_{m,m} \geq 1$ for all m. Induction on $k - m$ yields $a_{m,k} \geq 2^{k-m}$ for all k and m.

Assume that Λ is not block lacunary. Then, according to 7.1.4, Λ is standard and we have $\lim_{m \to \infty} \lambda_{m+1}/\lambda_m = 1$. Fix $\epsilon \in\,]0, 1[$ and find m_0 such that $\lambda_{m+1}/\lambda_m \leq 1 + \epsilon$ for all $m \geq m_0$. With the first part of Proposition 7.2.2 we obtain

$$2^{m-1} \leq \lambda_m \leq (1+\epsilon)^{m-m_0} \lambda_{m_0}$$

for all $m \geq m_0$ and hence

$$1 \leq \lim_{m \to \infty} \left(\frac{1+\epsilon}{2}\right)^{m-m_0} \left(\frac{1}{2}\right)^{m_0-1} \lambda_{m_0} = 0 \,,$$

a contradiction. $\square$

Proposition 7.2.3 *There exists a strictly absolutely monotone non-lacunary sequence Λ of integers.*

Proof. At first we observe that, if $\{\alpha_k\}_{k=1}^{\infty}$ is strictly absolutely monotone and $\{\beta_k\}_{k=1}^{\infty}$ is absolutely monotone then $\{\alpha_k + \beta_k\}_{k=1}^{\infty}$ is strictly absolutely monotone. This is a straightforward consequence of the definitions.

Now we use induction to introduce strictly absolutely monotone sequences $\{a_{m,j}\}_{j=1}^{\infty}$, $m = 1, 2, \ldots$, and indices $j_1 = 1 < j_2 < \ldots$. Put $a_{1,j} = 2^j$ and $j_1 = 1$.

If we have already $\{a_{m,j}\}_{j=1}^{\infty}$ and j_m for some m then let $j_{m+1} > j_m$ be such that

$$\frac{j_{m+1} - j_m}{j_{m+1} - 1 - j_m} \leq 1 + \frac{1}{2(m+1)} \,.$$

Let b be a positive integer with $a_{m,j_{m+1}}/b \leq 2^{-1}(m+1)^{-1}$. Put

$$a_{m+1,j} = \begin{cases} a_{m,j}, & j \leq j_m \\ a_{m,j} + b(j - j_m), & j > j_m \end{cases} \,.$$

Since $\underbrace{0, \ldots, 0}_{j_m \text{ times}}, b, 2b, 3b, \ldots$ is absolutely monotone, $\{a_{m+1,j}\}_{j=1}^{\infty}$ is strictly absolutely monotone and we obtain

$$\frac{a_{m+1,j_{m+1}}}{a_{m+1,j_{m+1}-1}} \leq 1 + \frac{1}{m+1} \,.$$

Finally, put $\lambda_j = a_{m,j}$ if $j_{m-1} < j \leq j_m$. It follows from the construction that $\Lambda = \{\lambda_j\}_{j=1}^{\infty}$ is a strictly absolutely monotone sequence of integers and we have

$$\lim_{m \to \infty} \frac{\lambda_{j_m}}{\lambda_{j_m - 1}} = \lim_{m \to \infty} \frac{a_{m,j_m}}{a_{m,j_m - 1}} = 1 \ .$$

Hence Λ is not lacunary.   □

7.3 Elementary Properties of Müntz Sequences and Polynomials

Now we focus on $M(\Lambda)$ instead of Λ. We want to discuss elementary properties of $M(\Lambda)$ where Λ satisfies some of the preceding conditions. We start with a technical lemma.

Lemma 7.3.1 *Let $g(t) = t^\lambda - t^\mu$, where $0 < \lambda < \mu$, and put $\rho = \mu/\lambda$. Then we obtain*

$$\|g\|_C = \nu(\rho) \cdot \left(1 - \frac{1}{\rho}\right) \quad with \quad \nu(\rho) = \rho^{\frac{1}{1-\rho}} \ .$$

$\nu(\rho)$ is a strictly increasing function on $]1, \infty[$ satisfying $1/e < \nu(\rho) < 1$ and $\lim_{\rho \to 1} \nu(\rho) = 1/e$. Moreover, $t_0 := (\lambda/\mu)^{1/(\mu - \lambda)}$ is the unique maximum point of $|g(t)|$.

Proof. It follows from simple calculus that g attains its unique maximum at t_0 and that $\|g\|_C = \nu(\rho)(1 - 1/\rho)$. We have

$$\frac{d \log \nu}{d\rho} = \frac{1/\rho - 1 + \log \rho}{(1 - \rho)^2}$$

and

$$\frac{d(1/\rho - 1 + \log \rho)}{d\rho} = \frac{1}{\rho}\left(1 - \frac{1}{\rho}\right) > 0 \quad \text{for } \rho > 1 \ .$$

Since $1/\rho - 1 + \log \rho = 0$ if $\rho = 1$ we obtain $\frac{d \log \nu}{d\rho} > 0$ and hence $\log \nu$ and ν are strictly increasing.   □

Lemma 7.3.1 has a number of consequences. At first we note

Proposition 7.3.2 *The Müntz sequences $\{t^{\lambda_k}\}_{k=1}^N$ and $\{t^{\mu_k}\}_{k=1}^N$ are isometrically equivalent in $C[0, 1]$ if and only if*

$$\frac{\lambda_{j+1}}{\lambda_j} = \frac{\mu_{j+1}}{\mu_j} \quad for \quad j = 1, 2, \dots, N - 1 \ .$$

Proof. The sufficiency of the condition for isometric equivalence follows easily by substituting $\tau = t^{\lambda_1}$ and $\tau = t^{\mu_1}$.

For the necessity assume that $\|t^{\lambda_1} - t^{\lambda_j}\|_C = \|t^{\mu_1} - t^{\mu_j}\|_C$. With Lemma 7.3.1 we conclude $\lambda_j/\lambda_1 = \mu_j/\mu_1$.   □

As a direct consequence of 2.7.2 we have

Proposition 7.3.3 *Let E be either L_p, for $1 \leq p < \infty$, or C. Then span $\{t^{\lambda_k}\}_{k=1}^{n} \subset E$ is a continuous function of $(\lambda_1, \ldots, \lambda_n)$ with respect to (the logarithm of) the Banach-Mazur distance and the ball opening Θ of subspaces in E.*

Using the last two propositions we obtain

Proposition 7.3.4 *Let Λ be a (a,q)-geometric sequence with $q > 1$. Then*

(a) $\{t^{\lambda_k}\}_{k=1}^{\infty}$ is isometrically equivalent in C to $\{t^{\lambda_k}\}_{k=l}^{\infty}$ for all $l = 1, 2, \ldots$,
(b) If Λ is finite, i.e. if $\Lambda = \{aq, aq^2, \ldots, aq^n\}$ for some n, then $M_{a,q} = [M(\Lambda)]_C$ is a continuous function of the parameters a and q with respect to the ball opening Θ as metric. Furthermore, for some $\epsilon > 0$ we find a Lipschitz constant $c(q, \epsilon)$ with

$$\Theta(M_{a,q}, M_{a,\tau}) \leq c(q, \epsilon) \cdot |q - \tau|, \qquad \tau \in [q - \epsilon, q + \epsilon] .$$

We finish this section with an estimate of the values of a Müntz polynomial which has only two summands.

Lemma 7.3.5 *There is a constant $\kappa > 0$ satisfying the following: Let $0 < \lambda < \mu$ and $g(t) = t^\lambda - t^\mu$. Then, for any $t \in [0,1]$, we have*

$$|g(t)| \leq \kappa t^{\lambda/2} \|g\|_C .$$

Proof. Put $\rho = \lambda/\mu$. Then, according to 7.3.1, we obtain

$$\|g\|_C = \rho^{\frac{\rho}{1-\rho}} (1 - \rho) .$$

Put $f(t) = t^{\lambda/2} - t^{\mu - \lambda/2}$. Then, with the preceding ρ, 7.3.1 implies

$$\|f\|_C = \left(\frac{\rho}{2 - \rho} \right)^{\frac{\rho}{2(1-\rho)}} \frac{2 - 2\rho}{2 - \rho} .$$

Put

$$\tau(s) = \left(\frac{1}{s(2-s)} \right)^{\frac{s}{2(1-s)}} \frac{2 - 2s}{2 - s}, \qquad s \in \,]0, 1[\, .$$

Then τ is continuous and we have

$$\lim_{s \to 0} \tau(s) = 1 \quad \text{and} \quad \lim_{s \to 1} \tau(s) = \lim_{x \to \infty} \tau \left(\frac{x}{1+x} \right) = 0 .$$

Hence there is a constant $\kappa > 0$ with $\tau(s) \leq \kappa$ for all $s \in \,]0, 1[$. We obtain, for any $t \in [0, 1]$,

$$|g(t)| \le t^{\lambda/2}\|f\|_C = t^{\lambda/2}\frac{\|f\|_C}{\|g\|_C}\|g\|_C$$
$$= t^{\lambda/2}\tau(\rho)\|g\|_C$$
$$\le \kappa t^{\lambda/2}\|g\|_C$$

$\square$

Now we turn to general Müntz polynomials with two summands.

Proposition 7.3.6 *Let* $p(t) = at^\lambda + bt^\mu$ *with* $0 < \lambda < \mu$. *Then for any* $t \in [0,1]$ *we have*
$$|p(t)| \le (2\kappa + 1)t^{\lambda/2}\|p\|_C .$$
where κ *is the constant of Lemma 7.3.5.*

Proof. Assume $\|p\|_C = 1$. We have
$$p(t) = at^\lambda + bt^\mu = a(t^\lambda - t^\mu) + (a+b)t^\mu .$$

Hence $p(1) = a+b$ and $|a+b| \le \|p\|_C = 1$. Put $g(t) = t^\lambda - t^\mu$. Then we obtain $|a| \cdot \|g\|_C = \|a(t^\lambda - t^\mu)\|_C \le 2$. Using 7.3.5 we see that, for any $t \in [0,1]$,
$$|p(t)| \le \kappa t^{\lambda/2}|a| \cdot \|g\|_C + t^\mu \le (2\kappa + 1)t^{\lambda/2} .$$

$\square$

Compare Proposition 7.3.6 with Corollary 6.1.3. There the constant depends on the given exponents λ_j while here κ is independent of λ and μ.

If t in the preceding proposition is small then $|p(t)|$ is small. In particular we obtain a lower estimate for
$$\min\{t_0 \in [0,1] \ : \ |p(t_0)| = \|p\|_C\}$$

In the next chapter we extend Proposition 7.3.6 to general Müntz polynomials.

7.4 Differences of Müntz Sequences

Lemma 7.3.1 implies that the elements of a Müntz sequence $M(\Lambda)$ have, in general, "bad" mutual disposition. As we have noted already, even if Λ has "large gaps", i.e. if $\sum_{k=1}^\infty 1/\lambda_k < \infty$, in general $M(\Lambda)$ is not a basis or uniformly minimal. (Lateron in 9.2 we shall see that $M(\Lambda)$ is a basis if and only if Λ is lacunary. This is also equivalent to the condition that Λ is uniformly minimal or separated.) For standard Λ the normalized elements of $M(\Lambda)$ are even closing (see 7.4.4, for the definition of closing see 2.2.1).

The situation does not improve if we go over to the sequence of differences. Again, we obtain a closing sequence in general. However, the geometry of differences of a Müntz sequence helps to understand more complicated phenomena such as the geometry of octants which we discuss in 7.5

Lemma 7.4.1 *Consider $\mu > \lambda > 0$ and $\tau > 0$ such that*

$$2\mu \le \lambda + \tau \qquad and \qquad \tau\lambda \le \mu^2 .$$

Then for the functions $g_1(t) = t^\lambda - t^\mu$, $g_2(t) = t^\mu - t^\tau$ and $\Delta(t) = g_1(t) - g_2(t) = t^\lambda - 2t^\mu + t^\tau$ we have $\|\Delta\|_C \le 5e^2\|g_1\|_C^2$.

Proof. By substituting $t^\lambda = s$ we can assume that $\lambda = 1$ and hence $\mu \ge 1$. So we deal with $g_1(t) = t - t^\mu$, $g_2(t) = t^\mu - t^\tau$ where $2\mu - 1 \le \tau \le \mu^2$. Consider two cases.

1. $\tau = 2\mu - 1$. Here, $\tau - 1 = 2(\mu - 1) = 2\alpha$ with $\alpha = \mu - 1$. Moreover,

$$\Delta(t) = t - 2t^\mu + t^{2\mu-1} = t(1 - t^\alpha)^2 ,$$

$$\Delta'(t) = 1 - 2\mu t^\alpha + (2\mu - 1)t^{2\alpha} \text{ and } \mu^2 - (2\mu - 1) = \alpha^2 .$$

So we have a unique point t_0 of maximum for Δ with $t_0^\alpha = 1/(2\mu - 1)$. Thus,

$$\|\Delta\|_C = (2\mu - 1)^{-1/\alpha}\left(1 - \frac{1}{2\mu - 1}\right)^2 \le \left(\frac{2(\mu - 1)}{2\mu - 1}\right)^2 \le \left(\frac{2(\mu - 1)}{\mu}\right)^2 .$$

By Lemma 7.3.1, $\|g_1\|_C \ge e^{-1}(1-1/\mu)$, so $\|\Delta\|_C \le (2(1-1/\mu))^2 \le 4e^2\|g_1\|_C^2$.

2. $\tau > 2\mu - 1$. Since $\tau \le \mu^2$ we have

$$\|t^{2\mu-1} - t^\tau\| \le 1 - \frac{2\mu - 1}{\tau} \le 1 - \frac{2\mu - 1}{\mu^2} = \left(1 - \frac{1}{\mu}\right)^2 \le e^2\|g_1\|_C^2 .$$

Now, using the previous case and triangle inequality we finish the proof of 7.4.1 ∎

As a consequence of Lemma 7.4.1 we obtain estimates for the inclination of the elements of $M(\Lambda)$ and their differences with respect to the sup-norm.

Theorem 7.4.2 *Let $\Lambda = \{\lambda_k\}_{k=1}^\infty$ and put $e_k = t^{\lambda_k}$, $g_k = e_{k+1} - e_k$, $k = 1, 2, \ldots$. Then we have, with respect to the sup-norm $\|\cdot\|_C$,*

(a) $(e_k, \widehat{e_{k+1}}) \sim 1 - \lambda_k/\lambda_{k+1}$ as $k \to \infty$. In particular, if Λ is a standard sequence, then $M(\Lambda)$ is closing.

(b) $(g_k, \widehat{g_{k+1}}) \le 10e^2(e_k, \widehat{e_{k+1}})$ provided that

$$2\lambda_{k+1} \le \lambda_k + \lambda_{k+2} \qquad and \qquad \lambda_k\lambda_{k+2} \le \lambda_{k+1}^2 .$$

If the latter conditions hold for all k and Λ is standard then $\{g_k\}_{k=1}^\infty$ is closing, too.

Proof. (a) is a direct consequence of Lemma 7.3.1. To prove (b) fix k. Then we have

$$(g_k,\widehat{g_{k+1}}) \le \frac{\|g_k - g_{k+1}\|_C}{\|g_k\|_C} = \frac{\|\Delta\|_C}{\|g_k\|_C} \le 5e^2\|g_k\|_C$$

and, with the notion of angle (see 1.2),

$$\|g_k\|_C = \|e_{k+1} - e_k\|_C = \varphi(e_k, e_{k+1}) \le 2(e_k,\widehat{e_{k+1}}) .$$

This finishes part (b) of the theorem.   □

For example $\Lambda = \{k^2\}_{k=1}^\infty$ satisfies the assumptions of Theorem 7.4.2 (b).
    Now we turn to the L_p-case for $1 \le p < \infty$. Recall that

$$\|t^\lambda\|_{L_p} = (\lambda p + 1)^{-1/p} .$$

Lemma 7.4.3 *For $0 < \lambda < \mu$ we obtain*

$$\frac{1}{2}\left(\frac{\lambda}{2^p\mu}\right)^{\frac{p\lambda/\mu+1/\mu}{p^2(1-\lambda/\mu)}} \le \|(\lambda p + 1)^{1/p}t^\lambda - (\mu p + 1)^{1/p}t^\mu\|_{L_p}$$

$$\le 1 - \left(\frac{\lambda}{\mu}\right)^{1/p} + p^{1/p}\left(1 - \frac{\lambda}{\mu}\right)^{1/p} .$$

Proof. We have, with $a = (\lambda p + 1)^{1/p}$ and $b = (\mu p + 1)^{1/p}$,

$$\|bt^\mu - at^\lambda\|_{L_p} \le (b-a)\|t^\mu\|_{L_p} + a\|t^\lambda - t^\mu\|_{L_p}$$

$$\le \left(1 - \frac{a}{b}\right) + a\left(\int_0^1 (t^\lambda - t^\mu)dt\right)^{1/p}$$

$$\le 1 - \frac{a}{b} + a\left(\frac{1}{\lambda+1} - \frac{1}{\mu+1}\right)^{1/p}$$

$$\le 1 - \left(\frac{\lambda}{\mu}\right)^{1/p} + \left(\frac{\lambda p+1}{\lambda+1}\right)^{1/p}\left(1 - \frac{\lambda}{\mu}\right)^{1/p}$$

$$\le 1 - \left(\frac{\lambda}{\mu}\right)^{1/p} + p^{1/p}\left(1 - \frac{\lambda}{\mu}\right)^{1/p} .$$

Here we used $|t^\lambda - t^\mu|^p \le (t^\lambda - t^\mu)$ and $(\lambda c+1)/(\mu c+1) \ge \lambda/\mu$ for any $c > 0$.
    Put $t_0 = (a/(2b))^{1/(\mu-\lambda)}$. Then $at^\lambda - bt^\mu \ge at^\lambda/2$ if $0 \le t \le t_0$. This implies

$$\|bt^\mu - at^\lambda\|_{L_p} \ge \frac{a}{2}\left(\int_0^{t_0} t^{\lambda p}dt\right)^{1/p}$$

$$= \frac{1}{2}t_0^{\lambda+1/p}$$

$$= \frac{1}{2}\left(\frac{\lambda p+1}{2^p(\mu p+1)}\right)^{\frac{1+\lambda p}{p^2(\mu-\lambda)}}$$

$$\ge \frac{1}{2}\left(\frac{\lambda}{2^p\mu}\right)^{\frac{p\lambda/\mu+1/\mu}{p^2(1-\lambda/\mu)}} .$$ □

We obtain, using

$$\widehat{(x,y)} \le ||x - y|| \le 2\widehat{(x,y)} \quad \text{if } ||x|| = 1 \text{ and } ||y|| = 1 ,$$

Theorem 7.4.4 *The following are equivalent*
(i) $\Lambda = \{\lambda_k\}_{k=1}^{\infty}$ *is standard*
(ii) Λ *is not block-lacunary*
(iii) $M(\Lambda)$ *is closing in* $C[0,1]$
(iv) $M(\Lambda)$ *is closing in* L_p *if* $1 \le p < \infty$

Proof. $(i) \Leftrightarrow (ii)$ follows from 7.1.2, $(i) \Leftrightarrow (iii)$ follows from 7.4.2 (a) and 7.3.1. $(i) \Leftrightarrow (iv)$ follows from 7.4.3. Note, if

$$\frac{1}{2} \left(\frac{\lambda_k}{2^p \lambda_{k+1}} \right)^{\frac{\lambda_k/\lambda_{k+1}+1/(p\lambda_{k+1})}{p(1-\lambda_k/\lambda_{k+1})}}$$

tends to 0, then $\lim_{k \to \infty} \lambda_k/\lambda_{k+1} = 1$. $\qquad \square$

7.5 The Inclination of Positive Octants of Müntz Sequences

The results of the preceding section allow us to get estimates of the inclinations of the positive octants for any Müntz sequence.

For any sequence $\bar{e} = \{e_k\}_{k=1}^{\infty}$ in a Banach space and integers n, m with $m \le n$ define the *positive (m,n)-octant* as

$$\Omega_{m,n}(\bar{e}) = \left\{ \sum_{k=m}^{n} \alpha_k e_k \; : \; \alpha_k \ge 0, \; k = m, \ldots, n \right\} ,$$

and *positive normed (m,n)-octant* as

$$\tilde{\Omega}_{m,n}(\bar{e}) = \left\{ x \; : \; x = \sum_{k=m}^{n} \alpha_k e_k, \; \alpha_k \ge 0, \; k = m, \ldots, n, \; ||x|| \le 1 \right\} .$$

Now we consider again an increasing sequence $\{\lambda_k\}_{k=1}^{\infty}$ of positive real numbers and put $e_k = t^{\lambda_k}$. The error of approximation of Müntz polynomials with positive coefficients can be estimated by the following theorem.

Theorem 7.5.1 *For all integers n and m with $0 \le m < n$ we have, with respect to the sup-norm on $[0,1]$,*

(a) $(\widehat{\tilde{\Omega}_{1,m}, \tilde{\Omega}_{m+1,n}}) = ||t^{\lambda_m} - t^{\lambda_{m+1}}||_C$

(b) $\dfrac{1}{2}||t^{\lambda_m} - t^{\lambda_{m+1}}||_C \le (\widehat{\Omega_{1,m}, \Omega_{m+1,n}}) \le ||t^{\lambda_m} - t^{\lambda_{m+1}}||_C$

Hence

$$(\Omega_{1,m}, \widehat{\Omega}_{m+1,n}) \sim 1 - \frac{\lambda_m}{\lambda_{m+1}} \text{ and } (\tilde{\Omega}_{1,m}, \widehat{\tilde{\Omega}}_{m+1,n}) \sim 1 - \frac{\lambda_m}{\lambda_{m+1}} \text{ as } m \to \infty$$

Proof. (a): Put $a = (\tilde{\Omega}_{1,m}, \widehat{\tilde{\Omega}}_{m+1,n})$. Take polynomials $p(t) = \sum_{k=1}^{m} \alpha_k t^{\lambda_k}$ and $q(t) = \sum_{k=m+1}^{n} \beta_k t^{\lambda_k}$ with

$$\alpha_k \geq 0, \ \beta_k \geq 0 \text{ for all } k, \ ||p||_C = \sum_{k=1}^{m} \alpha_k = 1, (\tilde{\Omega}_{1,m}, \widehat{\tilde{\Omega}}_{m+1,n}) = ||p - q||_C$$

and $||q||_C \leq 1$. Hence $\sum_{k=m+1}^{n} \beta_k \leq 1$. We obtain

$$||t^{\lambda_m} - t^{\lambda_{m+1}}||_C \geq a \geq p(t) - q(t) \geq t^{\lambda_m} - \sum_{k=m+1}^{n} \beta_k t^{\lambda_{m+1}} \geq t^{\lambda_m} - t^{\lambda_{m+1}} \geq 0$$

for all $t \in [0, 1]$. This implies $a = ||t^{\lambda_m} - t^{\lambda_{m+1}}||_C$.

(b): Let a be as in (a) and put $b = (\Omega_{1,m}, \widehat{\Omega}_{m+1,n})$. Take polynomials $p \in \Omega_{1,m}$ and $q \in \Omega_{m+1,n}$ such that $||p||_C = 1$ and $b = ||p - q||_C$. Then we have $||q||_C \leq b + 1$. Hence

$$b \geq \left\| p - \frac{q}{||q||_C} \right\|_C - ||q||_C + 1 \geq a - b$$

and we obtain $a/2 \leq b$. The right-hand inequality of (b) follows directly from the definitions.

The last assertion of Theorem 7.5.1 is a consequence of 7.4.2.   □

We also obtain the straightforward.

Proposition 7.5.2 *Let $E = C$ or $E = L_p$, $1 \leq p < \infty$. Then for any Müntz sequence $M(\Lambda)$ and any Müntz polynomial $p(t) = \sum_k \alpha_k t^{\lambda_k}$ we have*

$$||p||_E \leq \left\| \sum_k |\alpha_k| t^{\lambda_k} \right\|_E.$$

More on the Geometry of Müntz Sequences and Müntz Polynomials

In this chapter we continue the analysis of Müntz polynomials $f(t) = \sum_{k=1}^{n} \alpha_k t^{\lambda_k}$ where $0 < \lambda_1 < \lambda_2 < \dots$. Here we focus on three deep theorems. At first we give lower estimates for the maximum points of $|f(t)|$ in $[0,1]$. In this context we also compare the sup-norm of the function $f(t)t^{-\lambda_1/2}$ with the sup-norm of f. In the following section we consider the derivative f' of f and give estimates for $\|tf'(t)\|_C$ and $\|f'\|_C$ with the help of the sup-norm of f. Finally, we study the behaviour of a Müntz polynomial on a subset $\Omega \subset [0,1]$. This leads to versions of the Müntz theorem over Ω.

8.1 Lorentz-Saff-Varga-Type Theorems

Let $0 < \lambda_1 < \dots < \lambda_m$. Then a Müntz polynomial of the form $f(t) = \sum_{k=1}^{m} \alpha_k t^{\lambda_k}$ will be close to zero if t is close to zero. Hence the maximum points of $|f(t)|$ must lie somewhere in the right-hand part of $[0,1]$. We want to find concrete lower bounds for these points.

At first, it is helpful to consider a general (semi-)normed space E containing the continuous functions such that the operators T_ρ with $(T_\rho f)(t) = f(\rho t)$, $t \in [0,1]$, are bounded for $0 \le \rho \le 1$ and we have $\sup_{0 \le \rho \le 1} \|T_\rho\| < \infty$. Then we show

Lemma 8.1.1 *Any Müntz polynomial* $f(t) = \sum_{k=1}^{m} \alpha_k t^{\lambda_k}$ *satisfies*

$$\|T_\rho f\|_E \le 2 \left(\sum_{k=1}^{m} \rho^{\lambda_k \beta_k} \right) \sup_{0 \le \eta \le 1} \|T_\eta f\|_E \quad \text{for any} \quad \rho \in [0,1]$$

and any $\beta_k \ge 0$ *with* $\sum_{k=1}^{m} \beta_k = 1$.

Proof. We use induction on m. The case $m = 1$ is clear. Then assume that Lemma 8.1.1 holds for some m.

Finally, consider $f(t) = \sum_{k=1}^{m+1} \alpha_k t^{\lambda_k}$. At first we prove the assertion the sup-norm on $[0,1]$. Put $G = \text{span}\{t^{\lambda_{m+1}}\}$,

$$\tilde{\beta}_k = \frac{\beta_k}{\sum_{j=1}^m \beta_j}, \quad k = 1, \ldots, m, \quad \text{and} \quad \alpha = \sum_{j=1}^m \beta_j .$$

Then we have $1 - \alpha = \beta_{m+1}$. Now fix $\rho \in [0,1]$ and let $g \in G$ be such that

$$\|T_{\rho^\alpha} f + g\|_C = \inf_{\tilde{g} \in G} \|T_{\rho^\alpha} f + \tilde{g}\|_C .$$

We clearly have $\|g\|_C \leq 2\|f\|_C$. Using the induction hypothesis for $E = C/G$ we obtain, for any $t \in [0, \rho]$,

$$
\begin{aligned}
|f(\rho)| &\leq |T_{\rho^\alpha} f(\rho^{1-\alpha}) + g(\rho^{1-\alpha})| + |g(\rho^{1-\alpha})| \\
&\leq \inf_{\tilde{g} \in G} \|T_{\rho^\alpha} f + \tilde{g}\|_C + 2\|f\|_C \rho^{\lambda_{m+1}(1-\alpha)} \\
&= \|T_{\rho^\alpha} f + G\|_{C/G} + 2\|f\|_C \rho^{\lambda_{m+1}(1-\alpha)} \\
&\leq 2 \sum_{k=1}^m \rho^{\lambda_k \tilde{\beta}_k \alpha} \|f + G\|_{C/G} + 2\|f\|_C \rho^{\lambda_{m+1}(1-\alpha)} \\
&\leq 2 \sum_{k=1}^{m+1} \rho^{\lambda_k \beta_k} \|f\|_C
\end{aligned}
$$

Hence $|f(t)| \leq 2 \left(\sum_{k=1}^{m+1} \rho^{\lambda_k \beta_k} \right) \|f\|_C$ for $0 \leq t \leq \rho$.

Now let $\| \cdot \|_E$ be a general (semi-)norm. Fix $\Phi \in E^*$ of norm one and consider

$$h(\rho) := \Phi(T_\rho f) = \sum_{k=1}^{m+1} \alpha_k \rho^{\lambda_k} \Phi(t^{\lambda_k}).$$

We obtain, by what we just have proved already,

$$|\Phi(T_\rho f)| = |h(\rho)| \leq 2 \sum_{k=1}^{m+1} \rho^{\lambda_k \beta_k} \sup_{0 \leq \eta \leq 1} |\Phi(T_\eta f)| .$$

Taking the sup over all such Φ on both ends yields

$$\|T_\rho f\|_E \leq 2 \sum_{k=1}^{m+1} \rho^{\lambda_k \beta_k} \sup_{0 \leq \eta \leq 1} \|T_\eta f\|_E$$

which finishes the induction.   □

Now we turn to the sup-norm $\| \cdot \|_C$ on $[0,1]$. Reformulating 8.1.1 we obtain

Corollary 8.1.2 *Any Müntz polynomial* $f(t) = \sum_{k=1}^{m} \alpha_k t^{\lambda_k}$ *satisfies*

$$|f(t)| \leq 2 \left(\sum_{k=1}^{m} t^{\lambda_k \beta_k} \right) \|f\|_C \quad for \ any \quad t \in [0,1]$$

and any $\beta_k \geq 0$ *with* $\sum_{k=1}^{m} \beta_k = 1$.

For $m = 2$ and $\beta_1 = \beta_2 = 1/2$ we recover the estimate of Proposition 7.3.6.

Now we introduce the notion of Lorentz point.

Definition 8.1.3 *Let* $f \in C[0,1]$. *The set* $\{t : \|f\|_C = |f(t)|\}$ *is called the Chebychev set of* f *and is denoted by* $Ch(f)$.

The smallest element of $Ch(f)$ *is called Lorentz point and is denoted by* $t_L(f)$. *Hence*

$$t_L(f) = \min Ch(f) = \min\{t \ : \ |f(t)| = \|f\|_C\} \ .$$

For general Müntz polynomials we have the following theorem.

Theorem 8.1.4 *Let* $0 < \lambda_1 < \ldots < \lambda_m$ *and put* $f(t) = \sum_{k=1}^{m} \alpha_k t^{\lambda_k}$. *Then*

$$t_L(f) \geq \max \left(\left(\frac{1}{2m} \right)^{\sum_{k=1}^{m} 1/\lambda_k} \ , \ \left(\frac{1}{3} \right)^{\sum_{k=1}^{m} k/\lambda_k} \right)$$

Proof. Apply 8.1.2 with $\beta_k = (\lambda_k \sum_{j=1}^{m} 1/\lambda_j)^{-1}$. This yields

$$|f(t)| \leq 2m t^{1/\sum_{j=1}^{m} 1/\lambda_j} \|f\|_C$$

and hence $|f(t)| < \|f\|_C$ whenever $t < (2m)^{-\sum_{j=1}^{m} 1/\lambda_j}$.

If we take $\beta_k = k(\lambda_k \sum_{j=1}^{m} j/\lambda_j)^{-1}$ then 8.1.2 shows that

$$|f(t)| \leq \left(\sum_{k=1}^{m} t^{k/\sum_{j=1}^{m} j/\lambda_j} \right) \|f\|_C \ .$$

Hence, for $t \leq 3^{-\sum_{k=1}^{m} k/\lambda_k}$ we obtain

$$|f(t)| \leq 2 \left(\sum_{k=1}^{m} \frac{1}{3^k} \right) \|f\|_C < \|f\|_C \ .$$

Combining the two cases proves the theorem.   □

Observe that, for fixed m, the lower bound of $t_L(f)$ in 8.1.4 tends to 1 if λ_1 tends to ∞. For simultaneously large λ_1 and m the lower bound in Theorem 8.1.4 is not optimal. In [127] Saff and Varga showed

$$t_L(f) \geq \left(\frac{\lambda_1}{\lambda_1 + m} \right)^2$$

provided that the λ_k are integers. This is an extension of a result of Lorentz who proved the same estimate for polynomials of the form $\sum_{k=n}^{n+m} \alpha_k t^k$ [85]. The proofs for this estimate are more complicated than our proofs of 8.1.1 and 8.1.4. The preceding Theorem 8.1.4 suffices for our applications in Chap. 9.

Corollary 8.1.5 *Let* $f(t) = \sum_{k=1}^{m} \alpha_k t^{\lambda_k}$ *and* $g(t) = t^{-\lambda_1/2} f(t)$. *The we have*
  (a) $\|g\|_C \leq d\|f\|_C$
  (b) $|f(t)| \leq t^{\lambda_1/2} d\|f\|_C$ *for any* $t \in [0, 1]$
where
$$d = \min \left((2m)^{\sum_{k=1}^{m} \lambda_1/(2\lambda_k - \lambda_1)}, \; 3^{\sum_{k=1}^{m} k\lambda_1/(2\lambda_k - \lambda_1)} \right)$$

Proof. 8.1.4 applied to g yields $t_L(g) \geq d^{-2/\lambda_1}$. Since $g = t^{-\lambda_1/2} f$ this implies $\|g\|_C \leq d\|f\|_C$. Hence $|f(t)| = t^{\lambda_1/2}|g(t)| \leq t^{\lambda_1/2} d\|f\|_C$. $\square$

More generally, we also consider α-Lorentz points.

Definition 8.1.6 *For* $\alpha \in [0, 1]$ *and* $f \in C[0, 1]$ *we define the* α*-Lorentz point by*
$$t_{L,\alpha}(f) = \min\{t \in [0, 1] \; : \; |f(t)| \geq \alpha\|f\|_C\} .$$

Then Corollary 8.1.2 implies

Corollary 8.1.7 *Let* $0 < \lambda_1 < \ldots < \lambda_m$ *and* $0 \leq \alpha \leq 1$. *Then, for* $f(t) = \sum_{k=1}^{m} \alpha_k t^{\lambda_k}$, *we have*
$$t_{L,\alpha}(f) \geq \max \left(\left(\frac{\alpha}{2m}\right)^{\sum_{k=1}^{m} 1/\lambda_k}, \; \left(\frac{\alpha}{2+\alpha}\right)^{\sum_{k=1}^{m} k/\lambda_k} \right)$$

Proof. Again, apply 8.1.2 with $\beta_k = (\lambda_k \sum_{j=1}^{m} 1/\lambda_j)^{-1}$ and then with $\beta_k = k(\lambda_k \sum_{j=1}^{m} j/\lambda_j)^{-1}$. $\square$

8.2 The Newman Inequality

Here we want to estimate $\|f'\|_C$ for a Müntz polynomial f. We start with an estimate of $\|t f'(t)\|_C$.

Theorem 8.2.1 *Let* $0 \leq \lambda_1 < \lambda_2 < \ldots < \lambda_n$. *Then any Müntz polynomial* $f \in span\ \{t^{\lambda_k}\}_{k=1}^{n}$ *satisfies*
$$\|t f'(t)\|_C \leq 9 \left(\sum_{k=1}^{n} \lambda_k\right) \|f\|_C$$

Proof. (See [12]). Without loss of generality we can assume that $\sum_{k=1}^{n} \lambda_k = 1$. (Otherwise go over to $f(t^{1/\sum_{k=1}^{n} \lambda_k})$.)

Let $\tau : [0,\infty[\to]0,1]$ be defined by $\tau(s) = e^{-s}$. Instead of taking $f(t) = \sum_{k=1}^{n} \alpha_k t^{\lambda_k}$, $t \in [0,1]$, we consider $g(s) = (f \circ \tau)(s) = \sum_{k=1}^{n} \alpha_k e^{-\lambda_k s}$, $s \in [0,\infty[$. Let $\|\cdot\|$ be the sup-norm over $[0,\infty[$. Then we prove

$$\|g'\| \le 9\|g\| \tag{8.1}$$

(8.1) yields Proposition 8.2.1 since $g'(\tau^{-1}(t)) = -tf'(t)$, $t \in]0,1]$.

At first we use complex analysis to prove a number of claims. Define

$$B(z) = \prod_{j=1}^{n} \frac{z - \lambda_j}{z + \lambda_j} \quad \text{and} \quad U(s) = \frac{1}{2\pi i} \int_{\Gamma} \frac{z^2 e^{-sz}}{(1-z)B(z)} dz$$

where $\Gamma = \{z \in \mathbf{C} : |z - 1| = 1\}$. We claim

$$|B(z)| \ge \frac{1}{3} \quad \text{for all} \quad z \in \Gamma. \tag{8.2}$$

Indeed, for $z \in \Gamma$ the function $|(z - \lambda_j)/(z + \lambda_j)|$ attains its minimum at $z = 2$. Since $0 \le \lambda_j \le 1$ this implies

$$\left| \frac{z - \lambda_j}{z + \lambda_j} \right| \ge \frac{2 - \lambda_j}{2 + \lambda_j} = \frac{1 - \lambda_j/2}{1 + \lambda_j/2}.$$

Next we use the inequality

$$\left(\frac{1-x}{1+x} \right) \left(\frac{1-y}{1+y} \right) = \frac{1 - (x+y)}{1 + (x+y)} + \frac{2xy(x+y)}{(1+x)(1+y)(1+x+y)}$$
$$\ge \frac{1 - (x+y)}{1 + (x+y)}$$

for all $x, y \ge 0$.

For $z \in \Gamma$ we obtain

$$|B(z)| \ge \prod_{j=1}^{n} \frac{1 - \lambda_j/2}{1 + \lambda_j/2}$$
$$\ge \frac{1 - 2^{-1}\sum_{j=1}^{n} \lambda_j}{1 + 2^{-1}\sum_{j=1}^{n} \lambda_j} = \frac{1 - 1/2}{1 + 1/2} = \frac{1}{3}$$

Moreover we show

$$\int_{0}^{\infty} |U(s)| ds \le 6 \tag{8.3}$$

Indeed, for $z = 1 + e^{i\theta}$ we have $|z|^2 = 2 + 2\cos\theta$. Hence (8.2) and Fubini's theorem imply

$$\int_0^\infty |U(s)|ds = \int_0^\infty \frac{1}{2\pi} \left| \int_\Gamma \frac{z^2 e^{-sz}}{(1-z)B(z)} dz \right| ds$$

$$\leq \frac{3}{2\pi} \int_0^\infty \int_0^{2\pi} (2 + 2\cos\theta) e^{(-1-\cos\theta)s} d\theta ds$$

$$= \frac{3}{2\pi} \int_0^{2\pi} \frac{2 + 2\cos\theta}{1 + \cos\theta} d\theta = 6$$

Finally, we claim

$$\int_0^\infty e^{-\lambda_j s} U(s) ds = \lambda_j - 3 \tag{8.4}$$

Indeed, Fubini's theorem yields

$$\int_0^\infty e^{-\lambda_j s} U(s) ds = \int_0^\infty e^{-\lambda_j s} \frac{1}{2\pi i} \int_\Gamma \frac{z^2 e^{-sz}}{(1-z)B(z)} dz ds$$

$$= \frac{1}{2\pi i} \int_0^\infty \int_\Gamma \frac{z^2 e^{-(z+\lambda_j)s}}{(1-z)B(z)} dz ds$$

$$= \frac{1}{2\pi i} \int_\Gamma \frac{z^2}{(z+\lambda_j)(1-z)B(z)} dz$$

$$= \frac{1}{2\pi i} \int_{|z|=2} \left(\frac{z}{z+\lambda_j} \right) \left(\frac{z}{1-z} \right) \left(\frac{1}{B(z)} \right) dz$$

Observe that the preceding integrand has only poles at 1 and, by definition of $B(z)$, at λ_k for $k \neq j$. (This is the reason why we introduced $B(z)$.) All singularities lie in the interior of Γ. This explains the validity of the last equation.

To evaluate this integral we use the residue theorem. We have the following Laurent series expansions for $|z| > 1$:

$$\frac{z}{z+\lambda_j} = 1 - \frac{\lambda_j}{z} + \frac{\lambda_j^2}{z^2} + \dots$$

$$\frac{z}{1-z} = -1 - \frac{1}{z} - \frac{1}{z^2} - \dots$$

and

$$\frac{1}{B(z)} = \prod_{k=1}^n \frac{1 + \lambda_k/z}{1 - \lambda_k/z} = \prod_{k=1}^n \left(1 + 2\sum_{l=1}^\infty \left(\frac{\lambda_k}{z} \right)^l \right)$$

$$= 1 + 2 \left(\sum_{k=1}^n \lambda_k \right) /z + 2 \left(\sum_{k=1}^n \lambda_k \right)^2 /z^2 + \dots$$

$$= 1 + \frac{2}{z} + \frac{2}{z^2} + \dots$$

Hence

$$\left(\frac{z}{z+\lambda_j}\right)\left(\frac{z}{1-z}\right)\left(\frac{1}{B(z)}\right) = -1 + \frac{\lambda_j - 3}{z} + \dots$$

Since the value of the integral equals the coefficient of $1/z$ in the Laurent series expansion of the integrand we obtain (8.4). To prove the proposition consider $g(s) = \sum_{k=1}^n \alpha_k e^{-\lambda_k s}$. We have, for any $a > 0$,

$$\int_0^\infty g(s+a)U(s)ds = \int_0^\infty \sum_{k=1}^n \alpha_k e^{-\lambda_k s} e^{-a\lambda_k} U(s)ds$$

$$= \sum_{k=1}^n \alpha_k e^{-\lambda_k a} \int_0^\infty e^{-\lambda_k s} U(s)ds$$

$$= \sum_{k=1}^n \alpha_k (\lambda_k - 3) e^{-\lambda_k a}$$

$$= -g'(a) - 3g(a)$$

With (8.5) this implies

$$|g'(a)| \le 3|g(a)| + \left|\int_0^\infty g(s+a)U(s)ds\right| \le 3||g|| + 6||g|| \ .$$

Hence $||g'|| \le 9||g||$.   □

In Müntz spaces we obtain better estimates. We prove

Proposition 8.2.2 *Let $1 \le \lambda_1 < \lambda_2 < \dots$ such that $\sum_{k=1}^\infty 1/\lambda_k < \infty$. Then there is a constant $K > 0$ with*

$$||f'||_C \le K\left(\sum_{k=m}^n \lambda_k\right)||f||_C$$

whenever $f(t) = \sum_{k=m}^n \alpha_k t^{\lambda_k}$ (for any m and n).

Proof. Since $\sum_{k=1}^\infty 1/\lambda_k < \infty$ we can apply Corollary 6.1.3 to $g(t) = tf'(t)$. Therefore we obtain a constant d with $|g(t)| \le dt^{\lambda_1}||g||_C$ for all t. Hence $|f'(t)| \le dt^{\lambda_1 - 1}||g||_C$. Now Theorem 8.2.1 yields

$$|f'(t)| \le 9d\left(\sum_{k=m}^n \lambda_k\right)||f||_C \ .$$ □

A version of 8.2.2 remains true if we drop the assumption $\sum_{k=1}^\infty 1/\lambda_k < \infty$. In [12] a proof is outlined for the estimate

$$||f'||_C \le 18\left(\sum_{k=1}^n \lambda_k\right)||f||_C$$

whenever $f(t) = \sum_{k=1}^n \alpha_k t^{\lambda_k}$, $1 \le \lambda_1 < \dots < \lambda_n$ and $\inf_k(\lambda_{k+1} - \lambda_k) \ge 1$.

We note some consequences of Proposition 8.2.2

Proposition 8.2.3 *With the assumptions of Proposition 8.2.2 we have*

$$\|f\|_C \le K^{1/p} \left(1 + \sum_{k=m}^{n} \lambda_k\right)^{1/p} n^2 \|f\|_{L_p}$$

if $1 \le p < \infty$.

Proof. At first, assume that $p = 2^l$ for some integer l. Then f^p is a Müntz polynomial with n^{2^l} summands. We obtain, for any $x \in [0,1]$,

$$\int_0^1 |f(t)|^p dt \ge \left| \int_0^x f^p(t)dt \right| = |\tilde{f}(x)|,$$

where $\tilde{f}$ is a Müntz polynomial with $\tilde{f}' = f^p$. The Newman inequality, applied to $\tilde{f}$, yields

$$\|f\|_C^p = \|\tilde{f}'\|_C \le K \sum_{j_1,\dots,j_{2^l}} (\lambda_{j_1} + \dots + \lambda_{j_{2^l}} + 1)\|\tilde{f}\|_C$$

$$\le Kn^{2^l} \left(1 + \sum_{j=m}^{n} \lambda_j\right) \int_0^1 |f(t)|^p dt.$$

If p is arbitrary let l be such that $2^l \le p \le 2^{l+1}$. Assume without loss of generality $\|f\|_C = 1$. Then

$$\|f\|_{L_p} \le \|f\|_{L_{2^{l+1}}} \le 1 \quad \text{and} \quad |f(t)|^{2^{l+1}} \le |f(t)|^p \le 1$$

We obtain, by what we have proved already,

$$1 = \|f\|_C^p \le Kn^{2^{l+1}} \left(1 + \sum_{j=m}^{n} \lambda_j\right) \int_0^1 |f(t)|^{2^{l+1}} dt$$

$$\le Kn^{2p} \left(1 + \sum_{j=m}^{n} \lambda_j\right) \int_0^1 |f(t)|^p dt$$

Thus

$$\|f\|_C = 1 \le K^{1/p} n^2 \left(1 + \sum_{j=m}^{n} \lambda_j\right)^{1/p} \left(\int_0^1 |f(t)|^p dt\right)^{1/p}.$$

$\square$

We conclude this section with a corollary which we will apply in 9.3.

Corollary 8.2.4 *With the assumptions of Proposition 8.2.2 we have, if* $1 \le p < \infty$,

$$\left(\frac{1}{\rho}\int_0^\rho |f(s)|^p ds\right)^{1/p} = \|T_\rho f\|_{L_p}$$

$$\le \rho^{\lambda_m/2}\left(\frac{2+2\sum_{j=m}^n \lambda_j}{\lambda_m p+2}\right)^{1/p} K^{1/p}(2n+2-2m)^{2n+2-2m}\|f\|_{L_p}$$

Proof. Consider g with $g(t)t^{\lambda_m/2}(p\lambda_m/2+1)^{1/p} = f(t)$. Corollary 8.1.5 (a) implies

$$\|g\|_C \le \frac{(2n+2-2m)^{2n+2-2m}}{(p\lambda_m/2+1)^{1/p}}\|f\|_C .$$

Hence, by Proposition 8.2.3, with substitution $s = \rho t$,

$$\left(\frac{1}{\rho}\int_0^\rho |f(s)|^p ds\right)^{1/p} = \|T_\rho f\|_{L_p}$$

$$\le \|g\|_C \left(\frac{1}{\rho}\int_0^\rho \left(\lambda_m \frac{p}{2}+1\right)s^{\lambda_m p/2}ds\right)^{1/p}$$

$$= \rho^{\lambda_m/2}\|g\|_C$$

$$\le \rho^{\lambda_m/2}K^{1/p}(2n+2-2m)^{2n+2-2m}\left(\frac{1+\sum_{j=m}^n \lambda_j}{\lambda_m p/2+1}\right)^{1/p}\|f\|_{L_p}$$

$\square$

8.3 A Bernstein-Type Inequality in C

Here we discuss a somewhat different estimate for $\|f'\|_C$ where f is a Müntz polynomial. Now we assume that Λ is lacunary.

Theorem 8.3.1 *(See [12, 54, 59])* If $\Lambda = \{\lambda_k\}_{k=0}^\infty$ is lacunary with $\lambda_0 = 0$ then there exists a constant $c = c(\Lambda)$ such that, for any $f \in$ span $M(\Lambda)$, we have

$$|f'(t)| \le \frac{c}{1-t}\|f\|_C \quad \text{for all } t \in [0,1[.$$

Proof. Let $f \in$ span $M(\Lambda)$, $\|f\|_C = 1$, be such that

$$f(t) = a_0 + \sum_{j\ge 1} a_j t^{\lambda_j} .$$

By 6.3.2 $M(\Lambda)$ is a uniformly minimal system. So, by 2.2.2, there is a constant $c_1 > 0$, independent of f, such that, for all j, $|a_j| \le c_1$. Let $[\lambda]$ be the largest integer smaller than or equal to λ. Then we have

$$[\lambda_{j+1} - 1] - [\lambda_j - 1] + 1 \geq \lambda_{j+1} - \lambda_j \geq \left(1 - \frac{1}{q}\right)\lambda_{j+1}$$

where $q = \inf_{k \geq 1} \lambda_{k+1}/\lambda_k > 1$. Hence

$$\lambda_{j+1} t^{\lambda_{j+1}-1} \leq \lambda_{j+1} t^{[\lambda_{j+1}-1]} \leq \frac{q}{q-1}(t^{[\lambda_j-1]} + t^{[\lambda_j-1]+1} + \ldots + t^{[\lambda_{j+1}-1]}).$$

We conclude, with $c_2 = c_1 q/(q-1)$,

$$\begin{aligned}
|f'(t)| &\leq |\sum_j a_j \lambda_j t^{\lambda_j - 1}| \\
&\leq \sum_j |a_j| \lambda_j t^{\lambda_j - 1} \\
&\leq c_1 \sum_j \lambda_j t^{\lambda_j - 1} \\
&\leq c_2 \sum_{j=0}^{\infty} t^j = \frac{c_2}{1-t}
\end{aligned}$$

$\square$

8.4 Some Applications of the Clarkson-Erdös Theorem

To conclude this chapter we study the behaviour of a Müntz polynomial on a given subset of $[0,1]$. To this end, let $\Omega \subset [0,1]$ be compact and let E be one of the spaces $L_p[0,1]$, $1 \leq p < \infty$, or $C[0,1]$. Then we denote by $E(\Omega)$ the corresponding space $L_p(\Omega)$ (with respect to the Lebesgue measure on Ω) or $C(\Omega)$ (with respect to the sup-norm on Ω). Let $\overset{\circ}{\Omega}$ be the interior of Ω with respect to $[0,1]$. Moreover, consider $\lambda_0 = 0 < \lambda_1 < \lambda_2 < \ldots$ and put $\Lambda = \{\lambda_k\}_{k=0}^{\infty}$. At first we prove

Proposition 8.4.1 *Assume that $\sum_{k=1}^{\infty} 1/\lambda_k < \infty$ and $\inf_k(\lambda_{k+1} - \lambda_k) > 0$. If $1 \in \overset{\circ}{\Omega}$ then there is a constant $c > 0$ such that*

$$\|f\|_{E(\Omega)} \leq \|f\|_E \leq c\|f\|_{E(\Omega)}$$

for any $f \in [M(\Lambda)]_E$.

Proof. The first inequality is clear.

Assume we find $f_n \in [M(\Lambda)]_E$ with $\|f_n\|_E = 1$ and $\|f_n\|_{E(\Omega)} \leq 1/n$ for all n. By assumption there is $\epsilon > 0$ with $[1 - 2\epsilon, 1] \subset \Omega$. By 6.2.3 there are $\alpha_{n,k}$ with $f_n(t) = \sum_{k=0}^{\infty} \alpha_{n,k} t^{\lambda_k}$. According to 6.2.2 there is a constant $M > 0$ with

$$|\alpha_{n,k}| \leq M(1+\epsilon)^{\lambda_k}(\lambda_k + 1) \tag{8.5}$$

for all k and n (since $||t^{\lambda_k}||_E \geq ||t^{\lambda_k}||_{L_1} = (\lambda_k + 1)^{-1}$).

This means that the extended functions

$$\tilde{f}_n(z) = \sum_{k=0}^{\infty} \alpha_{n,k} z^{\lambda_k}$$

are uniformly bounded on $\{z \in \mathbf{C} : 0 < \text{Re } z \text{ and } |z| < 1 - \epsilon\}$. By Montel's theorem there is a subsequence of $\tilde{f}_n$ which converges uniformly on compact subsets of $\{z \in \mathbf{C} : 0 < \text{Re } z \text{ and } |z| < 1-\epsilon\}$ to an analytic function $\tilde{f}$. Since, by assumption, $\lim_{n\to\infty} ||f_n||_{E([1-2\epsilon,1-\epsilon])} = 0$ we obtain that $\tilde{f}|_{[1-2\epsilon,1-\epsilon]} = 0$ and hence $\tilde{f} = 0$. (8.5) also implies that, for any $\delta > 0$, there is a constant $d > 0$ such that $\sup_{0 \leq t \leq d} |f_n(t)| \leq \delta$. From the preceding, by perhaps going over to a suitable subsequence, we see that $\lim_{n\to\infty} ||f_n||_{E([d,1-2\epsilon])} = 0$ and $\lim_{n\to\infty} ||f_n||_{E([1-2\epsilon,1])} = 0$. We conclude $\limsup_{n\to\infty} ||f_n||_E \leq \delta$ which contradicts $||f_n||_E = 1$ if δ is small.

Hence there must be a constant $c > 0$ with

$$||f||_E \leq c||f||_{E(\Omega)} \quad \text{for all} \quad f \in [M(\Lambda)]_E. \quad \square$$

As a direct consequence of Proposition 8.4.1 we obtain a Müntz theorem on Ω.

Corollary 8.4.2 *Assume that* $\inf_k(\lambda_{k+1} - \lambda_k) > 0$. *Let* $\Omega \subset [0,1]$ *be compact such that* $\overset{\circ}{\Omega} \neq \emptyset$. *Then* $[M(\Lambda)]_{E(\Omega)} \neq E(\Omega)$ *if and only if* $\sum_{k=1}^{\infty} 1/\lambda_k < \infty$. *Moreover, if* $\sum_{k=1}^{\infty} 1/\lambda_k < \infty$ *then* $M(\Lambda)$ *is a complete minimal system in* $[M(\Lambda)]_{E(\Omega)}$. *In this case, for any* $\rho \in \overset{\circ}{\Omega}$ *and any* $\delta > 0$, *there is* $M > 0$ *such that*

$$|\alpha_k| \leq \left(\frac{1+\delta}{\rho}\right)^{\lambda_k} ||f||_{E(\Omega)} \quad \text{for all} \quad k \geq M$$

whenever $f(t) = \sum_k \alpha_k t^{\lambda_k} \in [M(\Lambda)]_{E(\Omega)}$.

Proof. Let $\sum_{k=1}^{\infty} 1/\lambda_k = \infty$. Then, by 6.1.5, we have $[M(\Lambda)]_E = E$. Hence we also have $[M(\Lambda)]_{E(\Omega)} = E(\Omega)$.

Now assume $\sum_{k=1}^{\infty} 1/\lambda_k < \infty$ and let $\rho \in \overset{\circ}{\Omega}$ such that $\rho > 0$. Observe that $[M(\Lambda)]_{E([0,\rho])}$ is isometric to $[M(\Lambda)]_E$. Find $\epsilon > 0$ such that $[\rho - \epsilon, \rho] \subset \Omega$. 8.4.1 implies that $[M(\Lambda)]_{E([0,\rho])}$ and hence $[M(\Lambda)]_E$ are isomorphic to $[M(\Lambda)]_{E([\rho-\epsilon,\rho])}$. Then, in view of 6.2.2, for every $\delta > 0$, there are constants $M > 0$ and $c > 0$ with $||\alpha_m t^{\lambda_m}||_{E([0,\rho])} \leq c(1+\delta)^{\lambda_m}||f||_{E(\Omega)}$ for every $f(t) = \sum_k \alpha_k t^{\lambda_k}$ and every $m \geq M$. This proves the last inequality of the corollary. In particular, $M(\Lambda)$ is a minimal complete system in $[M(\Lambda)]_{E(\Omega)}$. If $0 < \lambda$ and $\lambda \notin \Lambda$ then $M(\Lambda \cup \{\lambda\})$ is a minimal system in $[M(\Lambda \cup \{\lambda\})]_{E(\Omega)}$. Hence $[M(\Lambda)]_{E(\Omega)} \neq [M(\Lambda \cup \{\lambda\})]_{E(\Omega)}$ and therefore $[M(\Lambda)]_{E(\Omega)} \neq E(\Omega)$. $\square$

The if-part of the preceding Müntz theorem is certainly not true for general infinite $\Omega \subset [0,1]$. It is possible to construct a lacunary sequence Λ of exponents and a sequence $\Omega = \{t_n\}_{n=1}^{\infty} \subset [0,1]$ with $\lim_{n\to\infty} t_n = 1$ such that $[M(\Lambda)]_{C(\Omega)} = C(\Omega)$. For other examples see [1].

If one goes deeply into the theory of Chebychev approximation then one can improve the first statement of Corollary 8.4.2 considerably. Indeed, in [13] it was shown that this result remains true for the sup-norm on any $\Omega \subset [0,1]$ with positive Lebesgue measure even without the gap condition. The same holds for the following Remez-type inequality . Since we do not need these facts lateron we only present the easily proved cases.

Corollary 8.4.3 *Assume that* $\sum_{k=1}^{\infty} 1/\lambda_k < \infty$ *and* $\inf_k(\lambda_{k+1} - \lambda_k) > 0$. *Let* $\Omega \subset [0,1]$ *be compact such that* $\overset{\circ}{\Omega} \neq \emptyset$. *Then there is a constant* $c > 0$ *such that, for any* $t \in [0, \inf \Omega]$ *and any Müntz polynomial* f, *we have*

$$|f(t)| \leq c\|f\|_{E(\Omega)}$$

Proof. We find $\rho \in \overset{\circ}{\Omega}$ such that $\inf \Omega < \rho$. Put

$$\delta = \frac{1}{2}\left(\frac{\rho}{\inf \Omega} - 1\right)$$

(Of course we can assume $\inf \Omega > 0$.) For $f(t) = \sum_k \alpha_k t^{\lambda_k}$ Corollary 8.4.2 yields a constant $d > 0$, independent of f, such that

$$|\alpha_k| \leq d\left(\frac{1+\delta}{\rho}\right)^{\lambda_k} \|f\|_{E(\Omega)} \quad \text{for all} \quad k .$$

Hence

$$|f(t)| \leq \sum_k |\alpha_k|(\inf \Omega)^{\lambda_k}\|f\|_{E(\Omega)} \leq \sum_k \left(\frac{1}{2} + \frac{\inf \Omega}{2\rho}\right)^{\lambda_k} \|f\|_{E(\Omega)} .$$

Then the corollary follows with

$$c = \sum_{k=0}^{\infty} \left(\frac{1}{2} + \frac{\inf \Omega}{2\rho}\right)^{\lambda_k} .$$ $\square$

9

Operators of Finite Rank and Bases in Müntz Spaces

Now we turn to the Banach space $[M(\Lambda)]_E$ where $E = L_p$, $1 \leq p < \infty$, or $E = C$. This chapter is mainly devoted to the question whether every Müntz space $[M(\Lambda)]_{L_p}$ and $[M(\Lambda)]_C$ has a basis or other bounded approximation properties. At the same time, for special Λ, we clarify the isomorphic character of $[M(\Lambda)]_{L_p}$ and $[M(\Lambda)]_C$.

9.1 Special Finite Rank Operators

We start discussing the commuting bounded approximation property (CBAP) for $[M(\Lambda)]_E$ (see 5.1). Here let $E = L_p[0,1]$, $1 \leq p < \infty$, or $E = C[0,1]$.

Throughout this section let $\Lambda = \{\lambda_k\}_{k=1}^{\infty}$ be given with

$$0 < \lambda_1 < \lambda_2 < \ldots, \quad \inf_k(\lambda_{k+1} - \lambda_k) > 0, \quad \text{and} \quad \sum_{k=1}^{\infty} \frac{1}{\lambda_k} < \infty \qquad (9.1)$$

For $f \in [M(\Lambda)]_E$ and $0 < \rho < 1$ put, as before, $(T_\rho f)(t) = f(\rho t)$, $t \in [0,1]$. Then T_ρ is a contractive and compact operator on $[M(\Lambda)]_C$ which can be inferred for example from 6.2.2. Hence $[M(\Lambda)]_C$ has the metric approximation property which, by [17], implies that $[M(\Lambda)]_C$ has the CBAP. (This is also true for $[M(\Lambda)]_{L_p}$.) But here we want to discuss an explicit commuting approximating sequence of finite rank operators R_n which are defined on $[M(\Lambda)]_E$ for all preceding E.

To this end put

$$c_k = c_k(E) = \sup \left\{ |\alpha_k| \; : \; f(t) = \sum_j \alpha_j t^{\lambda_j} \in [M(\Lambda)]_E, \; \|f\|_E \leq 1 \right\}.$$

Fix a positive number μ. We introduce indices $N_0 = 0 < M_1 < N_1 < M_2 < N_2 < M_3 < \ldots$ and real numbers $0 < \rho_1 < \rho_2 < \ldots$ by induction: If N_{m-1} has been defined already then fix $M_m > N_{m-1}$ and take some $\rho_m < 1$ with

$$\rho_m \geq 1 - \left(\sum_{j=1}^{M_m} c_j \max(1, \lambda_j) \right)^{-1} \qquad (9.2)$$

Fix $N_m > M_m$ so large that

$$\sum_{k \geq N_m} c_k \rho_m^{\lambda_k} \leq \mu \qquad (9.3)$$

Such a choice of N_m is always possible. Indeed, let $\epsilon = 1 - \rho_m$. Then, by Theorem 6.2.2, there is $M > 0$ with

$$c_k \leq (1 + \epsilon)^{\lambda_k} \leq e^{\epsilon \lambda_k} \quad \text{for all } k \geq M .$$

We obtain

$$\sum_{k \geq M} c_k \rho_m^{\lambda_k} \leq \sum_{k \geq M} (1 + \epsilon)^{\lambda_k} (1 - \epsilon)^{\lambda_k} \leq \sum_{k \geq M} e^{-\epsilon^2 \lambda_k} .$$

Hence, for suitable $N_m \geq M$ the left-hand side of (9.3) is smaller than μ.

Now fix an index m. For a polynomial $f(t) = \sum_k \alpha_k t^{\lambda_k}$ put

$$(R_m f)(t) = \sum_{k=1}^{M_m} \alpha_k t^{\lambda_k} + \sum_{k=M_m+1}^{N_m} \alpha_k \rho_m^{\lambda_k} t^{\lambda_k}. \qquad (9.4)$$

Then R_m is a bounded linear operator which can be uniquely extended to a bounded operator on $[M(\Lambda)]_E$. This extension will also be denoted by R_m. Note that R_m depends on M_m, N_m and ρ_m (and hence on E).

Definition 9.1.1 M_m, N_m and ρ_m are called characteristic indices of R_m with respect to Λ and μ.

The R_m are indeed uniformly bounded. We denote the operator norm on $[M(\Lambda)]_E$ by $\| \cdot \|_E$, too.

Lemma 9.1.2 We have

$$\|R_m\|_E \leq \begin{cases} 1 + \rho_m^{-1/p} + \mu & \text{if } E = L_p \\ 2 + \mu & \text{if } E = C \end{cases}$$

Moreover, $R_m R_n = R_{\min(m,n)}$ if $m \neq n$, and $R_m \to \text{id}$ pointwise on $[M(\Lambda)]_E$. Finally, if $0 \leq a < b \leq 1$ then

$$\left(\int_a^b |R_m f|^p dt \right)^{1/p} \leq \left(\frac{1}{\rho_m} \int_{a\rho_m}^{b\rho_m} |f|^p dt \right)^{1/p} + (1+\mu)(b-a)^{1/p} \|f\|_{L_p}. \qquad (9.5)$$

Proof. Let $f(t) = \sum_k \alpha_k t^{\lambda_k} \in [M(\Lambda)]_E$. Consider T_{ρ_m} where ρ_m is the number of (9.2). We obtain $\|T_{\rho_m} f\|_{L_p} \le \rho_m^{-1/p} \|f\|_{L_p}$. According to (9.2) the choice of ρ_m yields

$$\sum_{k=1}^{M_m} |\rho_m^{\lambda_k} - 1| c_k \le \sum_{k=1}^{M_m} \max(1, \lambda_k) c_k (1 - \rho_m) \le 1 \ .$$

Hence, with (9.2) and (9.3), for $1 \le p \le \infty$,

$$\|R_m f - T_{\rho_m} f\|_{L_p} \le \sum_{k=1}^{M_m} |\rho_m^{\lambda_k} - 1| |\alpha_k| + \sum_{l=N_m+1}^{\infty} \rho_m^{\lambda_l} |\alpha_l|$$
$$\le (1 + \mu) \|f\|_{L_p}.$$

(Note that $\|R_m\|_{L_\infty} = \|R_m\|_C$.) This implies $\|R_m\|_{L_p} \le 1 + \rho_m^{-1/p} + \mu$ and $\|R_m\|_C \le 2 + \mu$. We obtain similarly, with

$$\frac{b^{\lambda_k p + 1} - a^{\lambda_k p + 1}}{\lambda_k p + 1} \le (b - a) \ ,$$

$$\left(\int_a^b |R_m f - T_{\rho_m} f|^p dt \right)^{1/p}$$
$$\le \sum_{k=1}^{M_m} |\rho_m^{\lambda_k} - 1| |\alpha_k| \left(\int_a^b t^{\lambda_k p} dt \right)^{1/p} + \sum_{l=N_m+1}^{\infty} |\alpha_l| \rho_m^{\lambda_l} \left(\int_a^b t^{\lambda_k p} dt \right)^{1/p}$$
$$\le (1 + \mu)(b - a)^{1/p} \|f\|_{L_p} \ .$$

Using the Minkowski inequality and the substitution $s = \rho_m t$ we derive (9.5). The remaining assertions follow directly from the definition of R_m. $\square$

In the following we also consider

$$[M(\Lambda)]_C^0 := \text{closed span}\{t^{\lambda_{k+1}} - t^{\lambda_k} : k = 1, 2, \ldots\}$$
$$= \{f \in [M(\Lambda)]_C : f(1) = 0\}$$

Since $[M(\Lambda)]_C^0$ is one-codimensional in $[M(\Lambda)]_C$ it will turn out that the Banach space nature (i.e. the isomorphic character) of both spaces is the same. (Indeed, according to 9.1.6, $[M(\Lambda)]_C^0$ contains a complemented isomorphic copy of c_0. This implies that $[M(\Lambda)]_C \oplus c_0 \sim [M(\Lambda)]_C^0 \oplus c_0 \sim [M(\Lambda)]_C^0$.)

We modify R_m to obtain a finite rank operator mapping into $[M(\Lambda)]_C^0$ as follows: For a Müntz polynomial $f(t) = \sum_k \alpha_k t^{\lambda_k}$ put

$$(\bar{R}_m f)(t) = \sum_{k=1}^{M_m} \alpha_k t^{\lambda_k} + \sum_{k=M_m+1}^{N_m-1} \alpha_k \rho_m^{\lambda_k} t^{\lambda_k} - \left(\sum_{k=1}^{M_m} \alpha_k + \sum_{k=M_m+1}^{N_m-1} \alpha_k \rho_m^{\lambda_k} \right) t^{\lambda_{N_m}}$$

Lemma 9.1.3 *We have* $\bar{R}_m\bar{R}_n = \bar{R}_{\min(m,n)}$ *if* $m \neq n$ *and* $\lim_{m\to\infty} \bar{R}_m f = f$ *for every* $f \in [M(\Lambda)]_C^0$. *Moreover,*

$$||\bar{R}_m||_C \leq 2||R_m||_C .$$

Proof. (9.4) and the definition of $\bar{R}_m$ yield

$$(\bar{R}_m f)(t) = (R_m f)(t) - (R_m f)(1) \cdot t^{\lambda_{N_m}} .$$

From this we derive the lemma.   □

Proposition 9.1.4 *Fix* $E = L_p$, $1 \leq p < \infty$ *or* $E = C$. *Then there are* d_1, $d_2 > 0$ *and indices* $m_1 < m_2 < \dots$ *(depending on* E*) such that*

$$d_1 \left(\sum_k ||(R_{m_{k+1}} - R_{m_k})f||_{L_p}^p \right)^{1/p} \leq ||f||_{L_p}$$

$$\leq d_2 \left(\sum_k ||(R_{m_{k+1}} - R_{m_k})f||_{L_p}^p \right)^{1/p} \quad \text{if } f \in [M(\Lambda)]_{L_p},$$

and

$$d_1 \sup_k ||(\bar{R}_{m_{k+1}} - \bar{R}_{m_k})f||_C \leq ||f||_C$$

$$\leq d_2 \sup_k ||(\bar{R}_{m_{k+1}} - \bar{R}_{m_k})f||_C \quad \text{if } f \in [M(\Lambda)]_C^0$$

Proof. Put $R_0 = 0$. We define $0 = t_0 < t_1 < t_2 < \dots < 1$ and integers $0 = m_0 < m_1 < m_2 < \dots$ by induction: Put $m_1 = 1$. Assume we have already t_{k-1} and m_k. Then find $t_k \in]t_{k-1}, 1[$ such that

$$||(R_{m_k}f)1_{[t_k,1]}||_E \leq 6^{-k-1}||f||_E \text{ for all } f \in [M(\Lambda)]_E \text{ if } E = L_p \quad (9.6)$$

or $||(\bar{R}_{m_k}f)1_{[t_k,1]}||_C \leq 6^{-k-1}||f||_C$ for all $f \in [M(\Lambda)]_C^0$ if $E = C$.

This is possible since we have

$$\dim (R_{m_k}[M(\Lambda)]_E) < \infty \quad \text{and} \quad \lim_{t\to 1} \int_t^1 |(R_{m_k}f)|^p(s)ds = 0 \quad \text{if} \quad E = L_p$$

and $\lim_{t\to 1} ||(\bar{R}_{m_k}f)1_{[t,1]}||_C = 0$ if $f \in [M(\Lambda)]_C^0$.

Then take $m_{k+1} > m_k$ such that

$$||((id - R_{m_{k+1}})f)1_{[0,t_k]}||_E \leq 6^{-k}||f||_E \text{ for all } f \in [M(\Lambda)]_E \text{ if } E = L_p \quad (9.7)$$

or $||((id - \bar{R}_{m_{k+1}})f)1_{[0,t_k]}||_C \leq 6^{-k}||f||_C$ for all $f \in [M(\Lambda)]_C^0$ if $E = C$. Again this is possible since, for $f(t) = \sum_j \alpha_j t^{\lambda_j}$, by definition of $R_{m_{k+1}}$, there is some M such that

$$((id - R_{m_{k+1}})f)(t) = \sum_{j>M} \alpha_j \gamma_j t^{\lambda_j}$$

for certain numbers $\gamma_j \in [0,1]$. (To be more precise, according to (9.4), $M = M_{m_{k+1}}$, $\gamma_j = 1 - \rho_{m_{k+1}}^{\lambda_j}$, if $M_{m_{k+1}} < j \leq N_{m_{k+1}}$, and $\gamma_j = 1$, if $N_{m_{k+1}} < j$.) Let $\epsilon > 0$ be such that $t_k + \epsilon < 1$. Using 6.2.2, for large enough M and hence m_{k+1}, we obtain $|\alpha_j| \leq (1-\epsilon)^{-\lambda_j} \|f\|_E$, if $j > M$. Thus we have

$$\|((id - R_{m_{k+1}})f)1_{[0,t_k]}\|_E \leq \left(\sum_{j>M} \frac{1}{(1-\epsilon)^{\lambda_j}} t_k^{\lambda_j} \right) \|f\|_E \quad \text{if} \quad E = L_p$$

and, similarly,

$$\|((id - \bar{R}_{m_{k+1}})f)1_{[0,t_k]}\|_C \leq \|((id - R_{m_{k+1}})f)1_{[0,t_k]}\|_C + |(R_{m_{k+1}}f)(1)|t_k^{\lambda_{N_{m_{k+1}}}}$$

$$\leq \left((2+\mu)t_k^{\lambda_{N_{m_{k+1}}}} + \sum_{j>M} \frac{1}{(1-\epsilon)^j} t_k^{\lambda_j} \right) \|f\|_C \quad \text{if} \quad f \in [M(\Lambda)]_C^0 .$$

We take m_{k+1} and $M = M_{m_{k+1}}$ so large that

$$(2+\mu)t_k^{\lambda_{N_{m_{k+1}}}} + \sum_{j>M} (1-\epsilon)^{-j} t_k^{\lambda_j} \leq 6^{-k} .$$

Now, with (9.6) and (9.7), we have

$$\|f1_{[t_{k-1},t_k]}\|_{L_p} \leq \|((R_{m_{k+1}} - R_{m_{k-1}})f)1_{[t_{k-1},t_k]}\|_{L_p} +$$
$$\|((R_{m_{k-1}})f)1_{[t_{k-1},t_k]}\|_{L_p} + \|((id - R_{m_{k+1}})f)1_{[t_{k-1},t_k]}\|_{L_p}$$
$$\leq \|((R_{m_{k+1}} - R_{m_{k-1}})f)1_{[t_{k-1},t_k]}\|_{L_p} + \frac{2}{6^k}\|f\|_{L_p}.$$

Summation yields, using $(a+b)^p \leq 2(a^p + b^p)$ for $a, b > 0$,

$$\int_0^1 |f|^p dt \leq 2 \sum_{k=1}^{\infty} \left(\int_{t_{k-1}}^{t_k} |(R_{m_{k+1}} - R_{m_{k-1}})f|^p dt + \frac{2^p}{6^{kp}}\|f\|_{L_p}^p \right) .$$

This implies, with the Minkowski inequality and the fact that $R_{m_{k+1}} - R_{m_{k-1}} = (R_{m_{k+1}} - R_{m_k}) + (R_{m_k} - R_{m_{k-1}})$,

$$\left(\frac{1}{2} - \frac{2}{5} \right) \|f\|_{L_p}^p \leq \sum_k \|(R_{m_{k+1}} - R_{m_{k-1}})f\|_{L_p}^p$$

and hence

$$\|f\|_{L_p} \le 10 \left(\sum_k \|(R_{m_{k+1}} - R_{m_{k-1}})f\|_{L_p}^p \right)^{1/p}$$

$$\le 20 \left(\sum_k \|(R_{m_k} - R_{m_{k-1}})f\|_{L_p}^p \right)^{1/p} \tag{9.8}$$

Similarly, for $E = C$, if $f \in [M(\Lambda)]_C^0$, we obtain,

$$\|f\|_C \le \sup_k \|f 1_{[t_{k-1}, t_k]}\|_C$$

$$\le 10 \sup_k \|(\bar{R}_{m_{k+1}} - \bar{R}_{m_{k-1}})f 1_{[t_{k-1}, t_k]}\|_C$$

$$\le 20 \sup_k \|(\bar{R}_{m_k} - \bar{R}_{m_{k-1}})f\|_C \tag{9.9}$$

Here we also have

$$\sup_k \|(\bar{R}_{m_k} - \bar{R}_{m_{k-1}})f\|_C \le 2(\sup_k \|\bar{R}_{m_k}\|_C)\|f\|_C .$$

It remains to show the left-hand inequality of the proposition for the case $E = L_p$. At first, (9.6) and (9.7) yield

$$\int_0^1 |(R_{m_{k+1}} - R_{m_k})f|^p dt \le 2 \int_{t_{k-1}}^{t_{k+1}} |(R_{m_{k+1}} - R_{m_k})f|^p dt + \frac{4}{6^{kp}}\|f\|_{L_p}^p \tag{9.10}$$

(9.5) implies

$$\int_{t_{k-1}}^{t_{k+1}} |(R_{m_{k+1}} - R_{m_k})f|^p dt \le 2 \int_{t_{k-1}}^{t_{k+1}} |R_{m_{k+1}}f|^p dt + 2 \int_{t_{k-1}}^{t_{k+1}} |R_{m_k}f|^p dt \le$$

$$\frac{4}{\rho_{m_k}} \int_{t_{k-1}\rho_{m_k}}^{t_{k+1}\rho_{m_k}} |f|^p dt + \frac{4}{\rho_{m_{k+1}}} \int_{t_{k-1}\rho_{m_{k+1}}}^{t_{k+1}\rho_{m_{k+1}}} |f|^p dt + 8(1+\mu)^p(t_{k+1} - t_{k-1})\|f\|_{L_p}^p$$

Since $\rho_1 < \rho_2 < \ldots$ and $t_1 < t_2 < \ldots$ the intervals $[t_k \rho_{m_k}, t_{k+1}\rho_{m_k}]$, $k = 1, 2, \ldots$, are mutually disjoint. Similarly the intervals $[t_k \rho_{m_{k+1}}, t_{k+1}\rho_{m_{k+1}}]$ are mutually disjoint. Hence, with (9.10),

$$\sum_k \|(R_{m_{k+1}} - R_{m_k})f\|_{L_p}^p \le \frac{32}{\rho_1}\|f\|_{L_p}^p + 32(1+\mu)^p\|f\|_{L_p}^p + 5\|f\|_{L_p}^p$$

$\square$

In the preceding proof, in (9.6) and (9.7), we did not use the fact that the R_m are uniformly bounded. The argument can be applied even in the case where all ρ_m are zero. Then we obtain similarly

Corollary 9.1.5 *Let* $E = L_p$, $1 \le p < \infty$. *Then there are constants* $d_1, d_2 > 0$, *indices* $m_1 < m_2 < \dots$ *and numbers* $0 = t_0 < t_1 < t_2 < \dots < 1$ *(depending on E) such that*

$$d_1 \left(\sum_k \int_{t_k}^{t_{k+1}} \left| \sum_{j=m_k+1}^{m_{k+2}} \alpha_j t^{\lambda_j} \right|^p dt \right)^{1/p} \le \|f\|_{L_p}$$

$$\le d_2 \left(\sum_k \int_{t_k}^{t_{k+1}} \left| \sum_{j=m_k+1}^{m_{k+2}} \alpha_j t^{\lambda_j} \right|^p dt \right)^{1/p}$$

if $f = \sum_k \alpha_k t^{\lambda_k} \in [M(\Lambda)]_{L_p}$.

Proof. Here put $R_m \sum_k \alpha_k t^{\lambda_k} = \sum_{k=1}^{m} \alpha_k t^{\lambda_k}$. Then the R_m are bounded although not uniformly bounded in general. (6) and (7) yield

$$\|(f - (R_{m_{k+2}} - R_{m_k})f) 1_{[t_k, t_{k+1}]} \|_{L_p} \le \frac{2}{6^{k+1}} \|f\|_{L_p}$$

for all $f \in [M(\Lambda)]_{L_p}$ from which the corollary follows.   □

Put $F_k = \mathrm{span}\, \{t^{m_k+1}, \dots, t^{m_k+2}\}$. Then Corollary 9.1.5 suggests that the F_k are the summands of an FDD of $[M(\Lambda)]_{L_p}$. This is, however, not true since F_k and F_{k+1} are "overlapping". If Λ is standard then for no choice of $\{m_k\}_{k=1}^{\infty}$ can $F_2, F_4, F_6, \dots$ be summands of an FDD. This is a consequence of 7.4.4.

With the notions of Chap. 5 we obtain

Theorem 9.1.6 *Let* $\Lambda = \{\lambda_k\}_{k=1}^{\infty}$ *satisfy*

$$\sum_{k=1}^{\infty} \frac{1}{\lambda_k} < \infty \qquad and \qquad \inf_k (\lambda_{k+1} - \lambda_k) > 0$$

Then the following hold.
(a) $[M(\Lambda)]_E$ *has the shrinking CBAP if* $E = C$ *or* $E = L_p$, $1 < p < \infty$.
(b) $[M(\Lambda)]_{L_1}$ *has a* w^*-*shrinking CBAP. In particular,* $[M(\Lambda)]_{L_1}$ *is isomorphic to a dual space.*
(c) $[M(\Lambda)]_E$ *is isomorphic to a subspace of* l_p, *if* $E = L_p$, $1 \le p < \infty$, *and of* c_0, *if* $E = C$.
(d) $[M(\Lambda)]_E$ *contains a complemented isomorphic copy of* l_p, *if* $E = L_p$, *and of* c_0, *if* $E = C$.

Proof. (a) and (b) follow directly from 9.1.4 and 5.2. (Observe that $[M(\Lambda)]_C^0 \oplus \mathbf{R} = [M(\Lambda)]_C$.) 9.1.4 yields furthermore finite dimensional subspaces $F_n \subset E$ such that $[M(\Lambda)]_E$ is isomorphic to a subspace of $(\sum_n \oplus F_n)_{(c_0)}$ if $E = C$ and to a subspace of $(\sum_n \oplus F_n)_{(l_p)}$ if $E = L_p$, $1 \le p < \infty$. We may even assume that $\sup_n d(F_n, l_p^{\dim F_n}) < \infty$ where $d(\cdot, \cdot)$ is the Banach-Mazur distance. Then $(\sum \oplus F_n)_{(l_p)}$ is isomorphic to l_p and $(\sum \oplus F_n)_{(c_0)}$ is isomorphic to

c_0. This implies (c). Finally, consider m_k as in 9.1.4 and find indices j_k such that $R_{m_k} t^{\lambda_{j_k}} = 0$ and $R_{m_{k+1}} t^{\lambda_{j_k}} = t^{\lambda_{j_k}}$ for each k. Put $G_k = \operatorname{span}\{t^{\lambda_{j_k}}\}$. Then, in view of 9.1.4, $(\sum \oplus G_k)_{(l_p)}$ is a complemented subspace of $[M(\Lambda)]_{L_p}$ which is isomorphic to l_p. Similarly, $(\sum \oplus G_k)_{(c_0)}$ is a complemented copy of c_0 in $[M(\Lambda)]_C$. $\square$

9.1.6 (c) was also obtained by different methods in [70], see also [21].

Corollary 9.1.7 *There is a subsequence $\Lambda' \subset \Lambda$ such that, for some finite dimensional subspaces $F_n \subset [M(\Lambda')]_E$, we have*

$$[M(\Lambda')]_E \text{ is isomorphic to } \begin{cases} (\sum_n \oplus F_n)_{(c_0)} & \text{if} \qquad E = C \\ (\sum_n \oplus F_n)_{(l_p)} & \text{if } E = L_p, \ 1 \le p < \infty \end{cases}$$

If Λ is non-quasilacunary then Λ' can be arranged to be non-quasilacunary, too.

Proof. Take R_m with characteristic indices M_m and N_m. Pick a suitable subsequence of the indices, say $m_1 < m_2 < \ldots$, such that $m_{k+1} - m_k > 2$ and

$$\Lambda' := \cup_{k=1}^{\infty} \{\lambda_j \ : \ N_{m_{2k-1}} \le j \le M_{m_{2k}}\}$$

is non-quasilacunary if Λ is non-quasilacunary. Then consider $R_{m_{2k}}|_{[M(\Lambda')]_E}$ and $\bar{R}_{m_{2k}}|_{[M(\Lambda')]_C^0}$. In view of (9.4) and the definition of $\bar{R}_m$, since

$$\Lambda' \cap \{\lambda_j \ : \ M_{m_{2k}} < j \le N_{m_{2k}}\} = \emptyset,$$

the $R_{m_{2k}}$ and $\bar{R}_{m_{2k}}$ are FDD-projections on $[M(\Lambda')]_E$ and $[M(\Lambda')]_C^0$, resp. Now, an application of Proposition 9.1.4 concludes the proof. $\square$

9.2 Lacunary Müntz Spaces

Here let us assume again, that E is the completion of C_0 under a given semi-norm $\| \cdot \|_E$ such that T_ρ with $(T_\rho f)(t) = f(\rho t)$, $f \in C_0$, is bounded for all $\rho \in [0,1]$ and that $\|t^{\lambda_k}\|_E \neq 0$ for all k.

Let $\Lambda = \{\lambda_k\}_{k=1}^{\infty}$. With 6.3.2 we obtain

Proposition 9.2.1 *Let $\sup_{\rho_0 \le \rho < 1} \|T_\rho\|_E < \infty$ for all $\rho_0 > 0$. Then $M(\Lambda)$ is a basis of $[M(\Lambda)]_E$ provided that Λ is lacunary.*

Proof. Let $\Lambda = \{\lambda_k\}_{k=1}^{\infty}$ be lacunary. We use the same argument as in 9.1. Put, for $f(t) = \sum_k \alpha_k t^{\lambda_k}$, $(P_n f)(t) = \sum_{k=1}^n \alpha_k t^{\lambda_k}$. Define

$$c_k = \sup \left\{ |\alpha_k| \cdot \|t^{\lambda_k}\|_E \ : \ f = \sum_j \alpha_j t^{\lambda_j} \in [M(\Lambda)]_E, \|f\|_E \le 1 \right\}$$

According to 6.3.2 we have $c := \sup_k c_k < \infty$. Since Λ is lacunary there is a constant $d > 1/\lambda_1$ such that $\sum_{j=1}^{n} c_j \max(1, \lambda_j) \leq d\lambda_n$ for all n. Put $\rho_n = 1 - 1/(d\lambda_n)$. Then we have

$$\sum_{j=1}^{n} |\rho_n^{\lambda_j} - 1| c_j \leq \sum_{j=1}^{n} \frac{c_j \max(1, \lambda_j)}{d\lambda_n} \leq 1 \quad \text{and}$$

$$\sum_{k=n+1}^{\infty} c_k \rho_n^{\lambda_k} \leq \sum_{k=n+1}^{\infty} c \exp\left(-\frac{\lambda_k}{d\lambda_n}\right) \leq b$$

for some b depending only on $\inf_k \lambda_{k+1}/\lambda_k$. For $f = \sum_k \alpha_k t^{\lambda_k} \in [M(\Lambda)]_E$ we obtain

$$\|P_n f - T_{\rho_n} f\|_E \leq \sum_{j=1}^{n} |\rho_n^{\lambda_j} - 1| \cdot |\alpha_j| \cdot \|t^{\lambda_j}\|_E + \sum_{k=n+1}^{\infty} \rho_n^{\lambda_k} |\alpha_k| \cdot \|t^{\lambda_k}\|_E$$

$$\leq \left(\sum_{j=1}^{n} \frac{c_j \max(\lambda_j, 1)}{d\lambda_n} + \sum_{k=n+1}^{\infty} c_k \rho_k^{\lambda_k}\right) \|f\|_E$$

$$\leq (1 + b)\|f\|_E$$

Hence $\|P_n\|_E \leq \|T_{\rho_n}\|_E + 1 + b$. Since $\limsup_{n \to \infty} \|T_{\rho_n}\|_E < \infty$ the P_n are uniformly bounded. Hence $M(\Lambda)$ is a basis. $\qquad\square$

Let γ be the basis constant of the basis we introduced in the preceding proof. Then we also have shown that γ depends only on $\inf_k \{\lambda_{k+1}/\lambda_k\}$ and $\sup_\rho \|T_\rho\|_E$.

The construction of the P_n in the preceding proof is similar to the construction of the R_m in Sect. 9.1 but not identical. Here we used different c_k.

Theorem 9.2.2 *Let $E = L_p$, $1 \leq p < \infty$ or $E = C$. Then the following are equivalent.*

(i) Λ is lacunary
(ii) $M(\Lambda)$ is separated
(iii) $M(\Lambda)$ is uniformly minimal
(iv) $M(\Lambda)$ is a basis of $[M(\Lambda)]_E$

Proof. $(iv) \Rightarrow (iii) \Rightarrow (ii)$ follows from the definitions. $(ii) \Rightarrow (i)$ is a consequence of 7.4.2 and 7.4.3 and, finally, $(i) \Rightarrow (iv)$ follows from 9.2.1. $\qquad\square$

If $E = C$ then 9.2.2 $(i) \Leftrightarrow (iv)$ can be rephrased as follows:

$$[M(\Lambda)]_C = \left\{\sum_{k=1}^{\infty} \alpha_k t^{\lambda_k} : \sum_{k=1}^{\infty} \alpha_k t^{\lambda_k} \text{ converges uniformly on } [0,1]\right\}$$

if and only if Λ is lacunary.

Recall that the summing basis $\{g_k\}_{k=1}^\infty$ in c consists of the elements $g_k = (\underbrace{0,\ldots,0}_{k-1},1,1,\ldots)$. This sequence is a conditional basis of c. We have

$$\left\| \sum_{k=1}^\infty \alpha_k g_k \right\|_c = \sup_n \left| \sum_{k=1}^n \alpha_k \right| .$$

Corollary 9.2.3 *(See [54]) If Λ is lacunary and $E = C$ then $M(\Lambda)$ is equivalent to the summing basis of c. In particular $[M(\Lambda)]_C$ is isomorphic to c and hence to c_0. Actually, $d([M(\Lambda)]_C, c) \leq \gamma$ where γ is the basis constant of $M(\Lambda)$ and d is the Banach-Mazur distance.*

Proof. We have

$$\sup_n \left| \sum_{k=1}^n \alpha_k \right| \leq \sup_n \left\| \sum_{k=1}^n \alpha_k t^{\lambda_k} \right\|_C$$

$$\leq \gamma \left\| \sum_{k=1}^\infty \alpha_k t^{\lambda_k} \right\|_C$$

if γ is the basis constant of $M(\Lambda)$. On the other hand, for any m,

$$\sum_{k=1}^m \alpha_k t^{\lambda_k} = \left(\sum_{k=1}^m \alpha_k \right) t^{\lambda_m} + \left(\sum_{k=1}^{m-1} \alpha_k \right) (t^{\lambda_{m-1}} - t^{\lambda_m}) + \ldots + \alpha_1 (t^{\lambda_1} - t^{\lambda_2})$$

and

$$0 \leq t^{\lambda_m}, \quad 0 \leq t^{\lambda_{k-1}} - t^{\lambda_k}, \quad t^{\lambda_m} + \sum_{k=2}^m (t^{\lambda_{k-1}} - t^{\lambda_k}) \leq 1 .$$

This implies

$$\left\| \sum_{k=1}^m \alpha_k t^{\lambda_k} \right\|_C \leq \sup_n \left| \sum_{k=1}^n \alpha_k \right|$$

and proves the corollary.   □

Similarly if $E = L_p$ and Λ is lacunary put

$$e_k(t) = \frac{t^{\lambda_k}}{\|t^{\lambda_k}\|_E} \quad \text{and} \quad \bar{e} = \{e_k\}_{k=1}^\infty .$$

Then $\bar{e}$ is equivalent to the unit vector basis of l_p. In particular $[M(\Lambda)]_{L_p}$ is isomorphic to l_p [54]. We shall prove this fact in 9.3.3 and 9.3.4.

9.3 Quasilacunary Müntz Spaces

Here we extend the main results of the preceding section. We want to show that quasilacunary Müntz spaces are isomorphic to l_p if $E = L_p$ and to c_0 if $E = C$. In particular these spaces have bases although, if Λ is not lacunary, the quasilacunary Müntz sequence $M(\Lambda)$ is not a basic sequence (in view of 9.2.2).

At first we consider the sup-norm case.

Theorem 9.3.1 *Let $E = C$. Assume that $\Lambda = \{\lambda_j\}_{j=1}^{\infty}$ is quasilacunary. Let $0 = n_0 < n_1 < \ldots$ be indices, N a positive integer and $q > 1$ such that $n_{m+1} - n_m \leq N$, $m = 0, 1, \ldots$ and $\inf_{m \geq 1} \lambda_{n_m+1}/\lambda_{n_m} \geq q$. Put $F_m = \text{span}\{t^{\lambda_{n_m}+1}, \ldots, t^{\lambda_{n_m+1}}\}$.*

Then there are constants $d_1, d_2 > 0$ such that, for all $f_m \in F_m$, we have

$$d_1 \left(\sup_m ||f_m - f_m(1)t^{\lambda_{n_m+1}}||_C + \sup_m \left| \sum_{k=1}^{m} f_k(1) \right| \right) \leq \left\| \sum_m f_m \right\|_C$$

$$\leq d_2 \left(\sup_m ||f_m - f_m(1)t^{\lambda_{n_m+1}}||_C + \sup_m \left| \sum_{k=1}^{m} f_k(1) \right| \right)$$

Proof. By perhaps substituting t by s^M for large enough M we see that we may assume without loss of generality $\lambda_1 \geq 1$. Let $g_m \in F_m$ be such that $g_m(1) = 0$ for all m. Put $\mu_m = \lambda_{n_m+1}/2$ and, moreover, put $t_m = 1 - 1/\mu_m$. Then let $t_m \leq t < t_{m+1}$. We have, in view of 8.1.5 (b),

$$|g_k(t)| \leq (2N)^{2N} t_{m+1}^{\mu_k} ||g_k||_C \leq (2N)^{2N} \exp\left(-\frac{\mu_k}{\mu_{m+1}} \right) ||g_k||_C .$$

Since $\sum_{j=n_m+1}^{n_{m+1}} \lambda_j \leq N\lambda_{n_{m+1}}$ for all m the mean-value theorem together with 8.2.2 implies

$$|g_j(t)| = |g_j(1) - g_j(t)| \leq \frac{||g_j'||_C}{\mu_m} \leq c_1 N \frac{\mu_{j+1}}{\mu_m} ||g_j||_C$$

for some universal constant c_1. Hence we have

$$\left| \sum_j g_j(t) \right| \leq \sum_{j=1}^{m-1} c_1 N \frac{\mu_{j+1}}{\mu_m} ||g_j||_C + \sum_{k=m}^{\infty} (2N)^{2N} \exp\left(-\frac{\mu_k}{\mu_{m+1}} \right) ||g_k||_C .$$

Since $\inf_j \mu_{j+1}/\mu_j \geq q$ we obtain, with 6.3.3,

$$c_2 \sup_j ||g_j||_C \leq \left\| \sum_j g_j \right\|_C \leq c_3 \sup_j ||g_j||_C$$

for some universal constants $c_2, c_3 > 0$.

For arbitrary $f_m \in F_m$ put $g_m(t) = f_m(t) - f_m(1)t^{\lambda_{n_m}+1}$. Then $g_m \in F_m$ and $g_m(1) = 0$. Define $P(\sum_m f_m) = \sum_m g_m$. Then we obtain, with 6.3.3,

$$\left\| P\left(\sum_m f_m\right) \right\|_C \leq c_3 \sup_m \|g_m\|_C$$

$$\leq 2c_3 \sup_m \|f_m\|_C$$

$$\leq \frac{2c_3}{d_1} \left\| \sum_m f_m \right\|_C$$

Hence P is a bounded projection. Moreover

$$(id - P)\left(\sum_m f_m\right) = \sum_m f_m(1)t^{\lambda_{n_m}+1} .$$

Thus, 9.2.3 implies

$$c_4 \sup_m \left| \sum_{k=1}^m f_k(1) \right| \leq \left\| (id - P) \sum_m f_m \right\|_C \leq c_5 \sup_m \left| \sum_{k=1}^m f_k(1) \right|$$

for some universal $c_4, c_5 > 0$. We conclude

$$\frac{\min(c_2, c_4)}{2 + 2\|P\|} \left(\sup_m \|g_m\|_C + \sup_m \left| \sum_{k=1}^m f_k(1) \right| \right) \leq \left\| \sum_m f_m \right\|_C$$

$$\leq \max(c_3, c_5) \left(\sup_m \|g_m\|_C + \sup_m \left| \sum_{k=1}^m f_k(1) \right| \right)$$

$\square$

Corollary 9.3.2 *If Λ is quasilacunary then $[M(\Lambda)]_C$ is isomorphic to c_0. In particular $[M(\Lambda)]_C$ has a basis.*

Proof. Let, with the notions of 9.2.3,

$$X = \left\{ \sum_{m=1}^\infty g_m \ : \ g_m \in F_m \text{ and } g_m(1) = 0 \text{ for all } m \right\}$$

and $Y = \text{closed span}\{t^{\lambda_{n_m}+1}\}_{m=1}^\infty$. Then, according to 9.2.3, Y is isomorphic to c_0. Moreover, by 9.3.1, X is isomorphic to $(\sum \oplus G_m)_{(c_0)}$ where $G_m = \{f \in F_m : f(1) = 0\}$. Since $\sup_m \dim G_m < \infty$ we obtain for the Banach-Mazur distances $\sup_m d(G_m, l_\infty^{k_m}) < \infty$ where $k_m = \dim G_m$. Hence X is isomorphic to $(\sum \oplus l_\infty^{k_m})_{(c_0)}$ which is isometrically isomorphic to c_0. Finally we have $[M(\Lambda)]_C = X \oplus Y$. Thus $[M(\Lambda)]_C$ is isomorphic to $c_0 \oplus c_0$ and hence to c_0. $\square$

Now we turn to the L_p-case.

Theorem 9.3.3 *Let* $E = L_p$, $1 \leq p < \infty$. *Assume that* $\Lambda = \{\lambda_j\}_{j=1}^{\infty}$ *is quasilacunary. Let* $0 = n_0 < n_1 < \ldots$ *be indices, N a positive integer and* $q > 1$ *such that* $n_{m+1} - n_m \leq N$, $m = 0, 1, \ldots$, *and* $\inf_{m \geq 1} \lambda_{n_m+1}/\lambda_{n_m} \geq q$. *Put* $F_m = \text{span}\{t^{\lambda_{n_m}+1}, \ldots, t^{\lambda_{n_m+1}}\}$.

Then there are constants $d_1, d_2 > 0$ *such that, for all* $f_m \in F_m$, *we have*

$$d_1 \left(\sum_m \|f_m\|_{L_p}^p \right)^{1/p} \leq \left\| \sum_m f_m \right\|_{L_p} \leq d_2 \left(\sum_m \|f_m\|_{L_p}^p \right)^{1/p}.$$

In particular, $[M(\Lambda)]_{L_p}$ *is isomorphic to* l_p *and* $[M(\Lambda)]_{L_p}$ *has a basis.*

Proof. By substituting t by s^M for large enough M we see that it is no loss of generality to assume $\lambda_1 \geq 1$.

We may assume that $\lambda_{n+1}/\lambda_n \leq q^2$ for all n. Indeed, otherwise if, for some $j \geq 2$, $q^j \leq \lambda_{n+1}/\lambda_n \leq q^{j+1}$, then take $\Lambda \cup \{q\lambda_n, q^2\lambda_n, \ldots, q^{j-1}\lambda_n\}$ instead of Λ. (So we add some blocks of block length 1.) Using induction, by enlarging Λ, we see that Λ can always be arranged to satisfy $\lambda_{n+1}/\lambda_n \leq q^2$ for all n in addition without distroying the quasilacunarity. In particular we can assume

$$q \leq \frac{\lambda_{n_{m+1}}}{\lambda_{n_m}} \leq q^{2N} \quad \text{for all } n \tag{9.11}$$

(If the inequalities of 9.3.3 hold for the extended Λ then certainly also for the given Λ.)

We can apply 8.2.3 and 8.2.4. 8.2.3 yields a constant $c_1 > 0$ with

$$\|f_m\|_C \lambda_{n_{m+1}}^{-1/p} \leq c_1 \|f_m\|_{L_p} \quad \text{for any } m \text{ and any } f_m \in F_m \tag{9.12}$$

For the following estimates we formally define $\lambda_0 = \lambda_{n_0} = 1$ which implies, since $\lambda_{n_1} \geq 1$ by assumption,

$$\frac{1}{\lambda_{n_0}} \leq 1 \quad \text{and} \quad 1 - \frac{1}{\lambda_{n_1}} \leq \frac{1}{\lambda_{n_0}} \tag{9.13}$$

Put $t_0 = 0$ and $t_k = 1 - 1/\lambda_{n_k}$, $k = 1, 2, \ldots$. Then we have, in view of (9.11) and (9.13),

$$\sup_{m \geq 1} \sum_{k=m+1}^{\infty} \left(\frac{\lambda_{n_m}}{\lambda_{n_{k-1}}} \right)^{1/p} t_k^{\lambda_{n_m}/2} \leq \sum_{l=0}^{\infty} q^{-l/p} < \infty$$

and

$$\sup_m \sum_{k=1}^{m-1} \left(\frac{\lambda_{n_m}}{\lambda_{n_{k-1}}}\right)^{1/p} t_k^{\lambda_{n_m}/2} \leq \sup_m \sum_{k=1}^{m-1} \left(\frac{\lambda_{n_m}}{\lambda_{n_{k-1}}}\right)^{1/p} \exp\left(-\frac{1}{2}\frac{\lambda_{n_m}}{\lambda_{n_k}}\right)$$

$$\leq \max(q^{2N}, \lambda_{n_1}) \sup_m \sum_{k=1}^{m-1} \left(\frac{\lambda_{n_m}}{\lambda_{n_k}}\right)^{1/p} \exp\left(-\frac{1}{2}\frac{\lambda_{n_m}}{\lambda_{n_k}}\right)$$

$$\leq \max\left(q^{2N}, \lambda_{n_1}\right) \int_1^\infty x^{1/p} \exp\left(-\frac{1}{2}x\right) dx$$

$$< \infty$$

This implies

$$\tau := \sup_{m\geq 1} \sum_{\substack{k=1\\k\neq m}}^\infty \left(\frac{\lambda_{n_m}}{\lambda_{n_{k-1}}}\right)^{1/p} t_k^{\lambda_{n_m}/2} < \infty \tag{9.14}$$

Similarly we obtain

$$\gamma := \sup_k \sum_{\substack{m=1\\m\neq k}}^\infty \left(\frac{\lambda_{n_m}}{\lambda_{n_{k-1}}}\right)^{1/p} t_k^{\lambda_{n_m}/2} < \infty.$$

Fix $f = \sum_k \alpha_k t^{\lambda_k} \in [M(\Lambda)]_E$. To prove the right-hand inequality of 9.3.3 it suffices to assume $f_0 = 0$. Put

$$g_m(t) = \sum_{k=n_m+1}^{n_{m+1}} \alpha_k t^{\lambda_k - \lambda_{n_m}/2} \quad \text{and} \quad f_m(t) = g_m(t) t^{\lambda_{n_m}/2}, \quad m = 1, 2, \dots .$$

Then 8.1.5 (b) implies $\|g_m\|_C \leq (2N)^{2N}\|f_m\|_C$ for all m. We have $t_k - t_{k-1} \leq 1/\lambda_{n_{k-1}}$ for all k including $k = 1$ in view of (9.13). We conclude, with the Hölder inequality and (9.12),

$$\sum_{k=1}^\infty \int_{t_{k-1}}^{t_k} |f - f_k|^p dt \leq \sum_{k=1}^\infty \left(\sum_{\substack{m=1\\m\neq k}}^\infty \|g_m\|_C t_k^{\lambda_{n_m}/2}\right)^p (t_k - t_{k-1})$$

$$\leq (2N)^{2Np} \sum_k \left(\sum_{m\neq k} \frac{\|f_m\|_C}{\lambda_{n_m}^{1/p}} \left(\frac{\lambda_{n_m}}{\lambda_{n_{k-1}}}\right)^{1/p} t_k^{\lambda_{n_m}/2}\right)^p$$

$$\leq (2N)^{2Np}\gamma^{p-1} \sum_k \sum_{m\neq k} \frac{\|f_m\|_C^p}{\lambda_{n_m}} \left(\frac{\lambda_{n_m}}{\lambda_{n_{k-1}}}\right)^{1/p} t_k^{\lambda_{n_m}/2}$$

$$\leq (2N)^{2Np}\gamma^{p-1} q^{2N}\tau \sum_m \frac{\|f_m\|_C^p}{\lambda_{n_{m+1}}}$$

$$\leq (2N)^{2Np}\gamma^{p-1} q^{2N} c_1^p \tau \sum_m \|f_m\|_{L_p}^p \tag{9.15}$$

(9.15) implies

$$\|f\|_{L_p} = \left(\sum_{k=1}^{\infty} \int_{t_{k-1}}^{t_k} |f(t)|^p dt\right)^{1/p} \leq d_2 \left(\sum_m \|f_m\|_{L_p}^p\right)^{1/p}$$

for some constant $d_2 > 0$.

To prove the left-hand inequality observe that (9.12) implies, for $m = 0, 1, 2, \ldots$ and $f_m \in F_m$ and any b with $0 \leq b \leq 1$,

$$\left(\int_b^1 |f_m|^p dt\right)^{1/p} \leq \|f_m\|_C (1-b)^{1/p} \leq c_1 (1-b)^{1/p} \lambda_{n_{m+1}}^{1/p} \|f_m\|_{L_p} .$$

8.2.4 provides us with a constant $c_2 > 0$ such that, for any m and $f_m \in F_m$ and any a with $0 \leq a \leq 1$ we have

$$\left(\int_0^a |f_m|^p dt\right)^{1/p} \leq a^{\lambda_{n_m}/2 + 1/p} c_2 \|f_m\|_{L_p} .$$

Fix $d \geq 1$ so large that

$$\left(1 - \frac{d}{\lambda_{n_m}}\right)^{\lambda_{n_m}/2 + 1/p} c_2 \leq \frac{1}{3}, \quad \text{if } \lambda_{n_m} > d, \quad \text{and} \quad \frac{c_1}{d^{1/p}} \leq \frac{1}{3} .$$

Put $u_m = \max(1 - d/\lambda_{n_m}, 0)$, $v_m = 1 - 1/(d\lambda_{n_{m+1}})$, $m = 1, 2 \ldots$, and $u_0 = v_0 = 0$. Then we obtain

$$\left(\int_0^{u_m} |f_m|^p dt\right)^{1/p} \leq \frac{1}{3}\|f_m\|_{L_p} \text{ and } \left(\int_{v_m}^1 |f_m|^p dt\right)^{1/p} \leq \frac{1}{3}\|f_m\|_{L_p} .$$

Hence

$$\frac{1}{3}\|f_m\|_{L_p} \leq \left(\int_{u_m}^{v_m} |f_m|^p dt\right)^{1/p} \tag{9.16}$$

Moreover, since Λ is quasilacunary, there is a positive integer j, independent of m such that $v_{m-j} \leq u_{m+j}$. Hence

$$\int_{u_m}^{v_m} |f_m|^p dt \leq \int_{u_m}^{u_{m+j}} |f_m|^p dt + \int_{v_{m-j}}^{v_m} |f_m|^p dt \tag{9.17}$$

It suffices to assume $f_m \neq 0$ for only finitely many m. In view of (9.17), Corollary 6.3.4 (applied $2j$-times) yields a constant $c > 0$ with

$$\left(\sum_{m=1}^{\infty} \int_{u_m}^{v_m} |f_m|^p dt\right)^{1/p} \leq 2jc\|\sum_{m=1}^{\infty} f_m\|_{L_p} .$$

With (9.16) we find a constant $d_1 > 0$ satisfying

$$d_1 \left(\sum_{m=0}^{\infty} \|f_m\|_{L_p}^p\right)^{1/p} \leq \|\sum_{m=0}^{\infty} f_m\|_{L_p} .$$

The inequalities of 9.3.3 show that $[M(\Lambda)]_E$ is isomorphic to $(\sum_m \oplus F_m)_{(l_p)}$. Put $k_m = \dim F_m$. Then we have $k_m \leq N$ for all m. Hence $\sup_m d(F_m, l_p^{k_m}) < \infty$ and $[M(\Lambda)]_E$ is isomorphic to $(\sum_m \oplus l_p^{k_m})_{(l_p)}$ which is l_p. $\square$

Put

$$e_k(t) = \frac{t^{\lambda_k}}{\|t^{\lambda_k}\|_E} \quad \text{and} \quad \bar{e} = \{e_k\}_{k=1}^\infty$$

We obtain in particular the complete isomorphic classification of $[M(\Lambda)]_{L_p}$ for lacunary Λ.

Corollary 9.3.4 *If $E = L_p$, $1 \leq p < \infty$, and Λ is lacunary then $\bar{e}$ is equivalent to the unit vector basis of l_p. In particular, $[M(\Lambda)]_{L_p}$ is isomorphic to l_p.*

Proof. Take 9.3.3 with $N = 1$. $\square$

We conclude this section with a number of open problems.

Problems 9.3.5 *(1) Does there exist a non-quasilacunary Müntz space without basis? What about $\Lambda = \{k^2\}_{k=1}^\infty$? (B.Shekhtman conjectured that $[M(\{k^2\})]_C$ does not have a basis.)*
(2) Let Λ be non-quasilacunary. Is it possible to find a non-quasilacunary subsequence $\Lambda' \subset \Lambda$ such that $[M(\Lambda')]_E$ is isomorphic to l_p, if $E = L_p$, and $[M(\Lambda')]_C$ is isomorphic to c_0?
(3) Let Λ be non-quasilacunary. Is it possible to find a non-quasilacunary subsequence $\Lambda' \subset \Lambda$ such that $[M(\Lambda')]_E$ has a basis? (See also 10.3)
(4) Let Λ be arbitrarily non-dense. Is it possible to find a non-dense sequence $\Lambda'' \supset \Lambda$ such that $[M(\Lambda'')]_E$ is isomorphic to l_p, if $E = L_p$, and $[M(\Lambda'')]_C$ is ismorphic to c_0?
(5) Let Λ be arbitrarily non-dense. Is it possible to find a non-dense sequence $\Lambda'' \supset \Lambda$ such that $[M(\Lambda'')]_E$ has a basis?

On the other hand it is not easy to find an example of a non-quasilacunary Müntz space where the existence of a basis can be shown explicitly. In 10.3 we shall construct a non-quasilacunary Müntz space with basis which is block-lacunary.

9.4 Non-Existence of Monotone Bases in Non-dense Müntz Subspaces of C

A sequence $\bar{e} = \{e_k\}_{k=1}^\infty$ in a Banach E is called *monotone* if

$$(L_{1,n}, \widehat{L_{n+1,\infty}}) = 1, \quad n = 1, 2, \ldots, \text{ where } L_{n,m} = \text{ closed span } \{e_k\}_{k=n}^m .$$

(see 2.3.4) If $\bar{e}$ is a basis then $\bar{e}$ is monotone if and only if the basis constant is equal to one.

For studying questions about monotonicity of sequences in C frequently the following elementary proposition is useful. Recall, for a function $f \in C[0,1]$ the subset of $[0,1]$ of all maximum points for $|f(t)|$ is denoted by $\mathrm{Ch}(f)$ (Definition 8.1.6)

Proposition 9.4.1 *For given functions $f, g \in C$ where $f \neq 0 \neq g$, the condition*

$$\widehat{(f,g)} < 1$$

holds provided that, for each $t \in \mathrm{Ch}(f)$, we have

$$\mathrm{sign}\, f(t) = \mathrm{sign}\, g(t).$$

Proof. Without loss of generality we can assume $\|f\|_C = 1$. Then, by definition, we have

$$\widehat{(f,g)} = \inf_{\|\alpha f\|_C = 1} \inf_{\beta} \|\alpha f - \beta g\|_C = \inf_{\beta} \|f - \beta g\|_C.$$

Put $A = \{t \in [0,1] : f(t) \cdot g(t) > 0\}$. Then $\mathrm{Ch}(f) \subset A$. A is open in $[0,1]$. In view of our assumptions there is $\epsilon \in\,]0,1[$ such that $B := \{t \in [0,1] : |f(t)| \geq 1 - \epsilon\} \subset A$. Since B is compact we find $\delta \in\,]0,1[$ such that $|g(t)| \geq \delta\|g\|$ for $t \in B$. Put $\beta = \epsilon/(2\|g\|_C)$. Then we obtain $\|f - \beta g\|_C \leq 1 - \delta \cdot \epsilon/2$. $\square$

Now we prove that a Müntz space does not have a monotone basis. We even obtain more:

Theorem 9.4.2 *Let $X = [M(\Lambda)]_C$ be a non-dense Müntz space where $\Lambda = \{\lambda_k\}_{k=1}^{\infty}$ satisfies the gap condition $\inf_k(\lambda_{k+1} - \lambda_k) > 0$. Then X does not possess a complete linearly independent sequence which is monotone. In particular, X does not have a monotone basis.*

Proof. (See [129], p.244.) Let $\{e_k\}_{k=1}^{\infty}$ be a complete linearly independent sequence in X and let $h \in X$ be a non-negative strictly increasing function on $[0,1]$, for example $h(t) = t^{\lambda_k}$ for some $\lambda_k \in \Lambda$. Consider two cases:

1. For some integer $n \geq 2$ we have $h \in L_{1,n}$. Take a function $f \in L_{1,n}$, $f \neq 0$, such that $f(1) = 0$. Let $t_0 = \max\{t : t \in \mathrm{Ch}(f)\}$. Then, obviously, $t_0 < 1$ and we can assume $f(t_0) > 0$. Since $t_0^{\lambda_k} \neq 0$ for all k and $L_{1,n}$ is finite dimensional we find a function $g \in L_{n+1,\infty}$ with $g(t_0) > 0$. Continuity yields α with $0 < \alpha < 1 - t_0$ such that $g(t) > 0$ for $t \in [t_0, t_0 + \alpha[$. Take $0 < \beta < \alpha$ such that $f(t) > 0$ if $t \in [t_0, t_0 + \beta[$. Put

$$\epsilon = \frac{\|f\|_C - \sup_{\beta + t_0 \leq t \leq 1} |f(t)|}{2\|h\|_C}.$$

Then $\epsilon > 0$ since t_0 was the largest maximum point of f. Hence for the function $f_\epsilon = f + \epsilon h$ we have $f_\epsilon(t) > 0$ if $t \in [t_0, t_0 + \beta[$ and $\mathrm{Ch}(f_\epsilon) \subset [t_0, t_0 + \beta[$.

To prove the latter relation observe that $f_\epsilon(t_0) = \|f\|_C + \epsilon h(t_0)$. Hence $|f_\epsilon(t)| < f_\epsilon(t_0)$ if $t < t_0$ since h is increasing and non-negative. On the other hand, if $t_0 + \beta \leq t_1$ then

$$\epsilon(h(t_1) - h(t_0)) \leq \epsilon \|h\|_C < \|f\|_C - \sup_{t_0+\beta \leq t \leq 1} |f(t)| \, .$$

Hence

$$\begin{aligned} |f_\epsilon(t_1)| &\leq \sup_{t_0+\beta \leq t \leq 1} |f(t)| + \epsilon h(t_1) \\ &< \|f\|_C + \epsilon h(t_0) = f_\epsilon(t_0) \\ &\leq \|f_\epsilon\|_C. \end{aligned}$$

This proves $\mathrm{Ch}(f_\epsilon) \subset [t_0, t_0 + \beta[$.

Using Proposition 9.4.1 yields $\widehat{(f_\epsilon, g)} < 1$ and hence $\widehat{(L_{1,n}, L_{n+1,\infty})} < 1$.

2. $h \notin L_{1,n}$ for all n. In view of the completeness of $\{e_k\}_{k=1}^\infty$ in X we can approximate h with any given exactness by elements of $L_{1,n}$ if n is suitably large. Then we can repeat the arguments of case 1 to obtain again $\widehat{(L_{1,n}, L_{n+1,\infty})} < 1$ for suitably large n. This completes the proof.   □

Conjecture. There exists an infinite dimensional Müntz space in C without monotone basic sequence.

It is even unknown whether there exists any infinite dimensional Banach space without monotone basic sequence.

9.5 Balanced Sequences in C

Here we prove again that a quasilacunary Müntz space $[M(\Lambda)]_C$ has a basis. This time we use a slightly different approach and, in this context, we discuss some general properties of sequences in quasilacunary Müntz spaces which are of independent interest.

Definition 9.5.1 *Consider the sequences* $\bar{e} = \{e_k\}_{k=1}^\infty \subset C[0,1]$ *and* $\bar{t} = \{t_k\}_{k=1}^\infty \subset [0,1]$ *where* $t_1 < t_2 < \ldots$ *and* $\lim_{k \to \infty} t_k = 1$.
(i) Let $M < \infty$. *Then* $\bar{e}$ *is called M-restricted by* $\bar{t}$ *if, for each* k, *we have*

$$\sum_{j=k+1}^\infty \max_{0 \leq t \leq t_k} |e_j(t)| \leq M \, .$$

(ii) Let $0 < m \leq 1$. *Then* $\bar{e}$ *is called m-deflected by* $\bar{t}$ *if, for each* k *and* $f \in$ *span*$\{e_j\}_{j=1}^k$ *with* $\|f\|_C = 1$, *we have*

$$\max_{0 \leq t \leq t_k} |f(t)| \geq m \, .$$

(iii) $\bar{e}$ *is called* balanced *if* $\bar{e}$ *is* M-restricted *and* m-deflected *by* $\bar{t}$ *for some* $\bar{t}, M$ *and* m.

At first we prove a general theorem about balanced sequences in C.

Theorem 9.5.2 *Let $\bar{e} = \{e_k\}_{k=1}^{\infty} \subset C[0,1]$ be a normalized balanced sequence. Then the following are equivalent*

    (i) $\bar{e}$ is uniformly minimal in C

    (ii) $\bar{e}$ is basic in C

Proof. Only "$(i) \Rightarrow (ii)$" requires a proof. So assume that $\bar{e}$ is ρ-minimal for some $\rho > 0$ (see 1.2.6.) and M-restricted and m-deflected by $\bar{t}$ for some M, m and $\bar{t}$. Fix k and let $f = \sum_{j=1}^{k} \alpha_j e_j$ be such that $\|f\|_C = 1$. Furthermore, let $g = \sum_{j=k+1}^{\infty} \alpha_j e_j$. Put $\alpha = m(M+\rho)^{-1}$. Then we have $m - \alpha M = \alpha \rho$. To estimate $\|f + g\|_C$ we consider two cases:

1. $|\alpha_j| \leq \alpha$ for all j. Then take $s_k \in [0, t_k]$ such that $|f(s_k)| = \max_{0 \leq t \leq t_k} |f(t)|$. We obtain

$$\|f + g\|_C \geq |f(s_k) + g(s_k)| \geq |f(s_k)| - |g(s_k)|$$

$$\geq m - \sum_{j=k+1}^{\infty} |\alpha_j| \cdot |e_j(s_k)| \geq m - \alpha M = \alpha \rho.$$

2. $|\alpha_i| > \alpha$ for some i. Since $\bar{e}$ is ρ-minimal we have

$$\|f + g\|_C = \|\alpha_i e_i + \sum_{j \neq i} \alpha_j e_j\|_C \geq \alpha \rho.$$

In both cases we obtain $\|f + g\|_C \geq m\rho/(M + \rho)$. By 1.2.3, $\bar{e}$ is a basis with index $\gamma(\bar{e}) \geq m\rho/(M + \rho)$. $\qquad\square$

Now, we turn to Müntz sequences. We consider the quasilacunary sequence $\{\lambda_k\}_{k=1}^{\infty}$ such that

$$\frac{\lambda_{n_{k+1}}}{\lambda_{n_k}} \geq \beta \quad \text{for some } \beta > 1 \quad \text{and} \quad s := \sup_k (n_{k+1} - n_k) < \infty$$

Put $n_0 = 0$. Hence we obtain

$$\sum_{j=n_{m-1}+1}^{n_m} \lambda_j \leq s\lambda_{n_m}.$$

We retain the notion of n_k, s and β in the following.

Proposition 9.5.3 *Let $\{\lambda_k\}_{k=1}^{\infty}$ be a quasilacunary sequence. Put $p_k(t) = \sum_{j=n_{k-1}+1}^{n_k} \alpha_{k,j} t^{\lambda_j}$ where $\|p_k\|_C = 1$, $k = 1, 2, \ldots$. Then, there is a universal constant $d > 0$ such that, with $t_k = 1 - 1/(ds\lambda_{n_k})$, the sequence $\{p_k\}_{k=1}^{\infty}$ is $1/2$-deflected in C by $\{t_k\}_{k=1}^{\infty}$.*

Proof. By the Newman inequality 8.2.2, in view of $\|p_k\|_C = 1$, we have

$$\|p_k'\|_C \leq c \sum_{j=n_{k-1}+1}^{n_k} \lambda_j \leq c\lambda_{n_k} s$$

for some universal constant c. (We may assume that $\lambda_1 \geq 1$.) Put $t_k = 1 - 1/(2cs\lambda_{n_k})$. Let $u_k \in [0, 1]$ be such that $|p_k(u_k)| = 1$. Fix $\epsilon > 0$ such that

$$u_k - t_k \leq \epsilon \leq \min(u_k, 1 - t_k) .$$

Then $\epsilon \leq 1/(2cs\lambda_{n_k})$. The mean value theorem yields

$$|p_k(u_k - \epsilon)| \geq 1 - \|p_k'\|_C \cdot \epsilon \geq 1 - cs\lambda_{n_k}\epsilon \geq \frac{1}{2} .$$

Since $t_k \geq u_k - \epsilon$ we obtain

$$\max_{t \in [0, t_k]} |p_k(t)| \geq |p_k(u_k - \epsilon)| \geq \frac{1}{2} .$$

This proves Proposition 9.5.3 with $d = 2c$. □

Now we come to the main result of this section.

Theorem 9.5.4 *Assume that $\Lambda = \{\lambda_j\}_{j=1}^{\infty}$ is a quasilacunary sequence. Consider $\bar{p} = \{p_j\}_{j=1}^{\infty}$ where $p_j \in \text{span} \{t^{\lambda_{n_j}+1}, \ldots, t^{\lambda_{n_{j+1}}}\}$ and $\|p_j\|_C = 1$ for all j.*

Then $\bar{p}$ is a basic sequence in $[M(\Lambda)]_C$ if and only if $\bar{p}$ is uniformly minimal.

Proof. In view of 9.5.2 we need to show that $\bar{p}$ is a balanced sequence. By 9.5.3 it suffices to prove that $\bar{p}$ is M-restricted by $\bar{t}$ for

$$t_j = 1 - \frac{1}{2d\lambda_{n_j}}$$

and some $M < \infty$ where d is the constant of 9.5.3. We obtain, for any $l \geq j+1$,

$$t_j^{\lambda_{n_l}/2} \leq \left(1 - \frac{1}{2d\lambda_{n_j}}\right)^{\lambda_{n_l}/2} \leq \exp\left(-\frac{1}{4d}\beta^{l-j}\right) .$$

In view of 8.1.5 we have $|p_l(t)| \leq Kt^{\lambda_{n_l}/2}$ for all $t \in [0, 1]$ and some universal constant K since

$$\sum_{k=2}^{n_{j+1}-n_j} \frac{\lambda_{n_j}+1}{2\lambda_{k+n_j} - \lambda_{n_j}+1} \leq s .$$

Then we conclude

$$\sum_{l=j+1}^{\infty} \max_{t \in [0, t_j]} |p_l(t)| \leq K \sum_{l=j+1}^{\infty} t_j^{\lambda_{n_l}/2}$$

$$\leq K \sum_{l=j+1}^{\infty} \exp\left(-\frac{1}{4d}\beta^{l-j}\right) =: M < \infty$$

which shows that $M(\Lambda)$ is M-restricted by $\bar{t}$. □

Combining 9.5.4 with 6.3.3 we obtain again that $[M(\Lambda)]_C$ has a basis.

10

Projection Types and the Isomorphism Problem for Müntz Spaces

This chapter is mainly dedicated to the negative solution of the following

Problem of isomorphisms. Are all sparse infinite dimensional Müntz spaces pairwise isomorphic?

We give a negative answer for Müntz spaces in C which follows from the fact that there are Müntz spaces having C-projection type and Müntz spaces with L-projection type (see 1.6). This negative answer was recently obtained by the first named author and B.Shekhtman during discussions of the problems of geometry of Müntz spaces.

We also show that there is an uncountable number of pairwise non-isometric Müntz spaces in C.

10.1 The Projection Function of Müntz Spaces

At first we consider the geometric properties of Müntz spaces and Müntz sequences in C. We begin with the lacunary case. As a consequence of 9.2.3 and 1.5.1 we obtain

Theorem 10.1.1 *In* $C[0,1]$ *every lacunary Müntz space is isomorphic to* c_0. *Hence lacunary Müntz spaces in* C *have the C-projection type.*

We shall see in the next section that there are also Müntz spaces in C of L-projection type. Müntz spaces in L_p, if $1 \leq p < \infty$, can only have L-projection type. This is an easy consequence of Theorem 9.1.6 (d).

Recall, $\lambda(A)$ is the absolute projection constant of a finite dimensional Banach space A and $\lambda(A, B)$ is the relative projection constant if A is a subspace of B (see 1.6). $d(\cdot, \cdot)$ is the Banach-Mazur distance.

Theorem 10.1.2 *Let* $\Lambda = \{\lambda_k\}_{k=1}^{\infty}$ *satisfy the gap condition* $\inf_k(\lambda_{k+1} - \lambda_k) > 0$. *Then* $X = [M(\Lambda)]_{L_p}$, *for* $1 \leq p < \infty$, *has L-projection type.*

Proof. 9.1.6 (d) yields a complemented subspace $Y \subset X$ such that $d(Y, l_p) < \infty$. Let $Y_n \subset Y$ correspond to the span of the first n elements of the unit vector basis in l_p. Then $\sup_n d(Y_n, l_p^n) < \infty$ and $\sup_n \lambda(Y_n, X) < \infty$. On the other hand, we have $\sup_n \lambda(Y_n) = \infty$ by 1.5.2.   □

10.2 Some Müntz Spaces of L-Projection Type

For finite Müntz sequences in $C[0, 1]$ we need the following fact.

Theorem 10.2.1 (Almost Orthogonality) *Let m , n be positive integers, $q_1, q_2 > 1$ and $a > 0$. Then, for any $\epsilon > 0$ there is an integer-valued function $N = N(\epsilon, a, m, n, q_1, q_2) > 0$ such that, for all numbers $\mu_1, \ldots, \mu_m$ with $\mu_1 < \ldots < \mu_m \leq a$ and $\inf_j \mu_{j+1}/\mu_j \geq q_1$ and all numbers $\nu_1, \ldots, \nu_n$ with $N < \nu_1 < \nu_2 < \ldots < \nu_n$ and $\inf_k \nu_{k+1}/\nu_k \geq q_2$, we have*

$$\left(\operatorname{span} \{t^{\mu_j}\}_{j=1}^m, \ \widehat{\operatorname{span} \{t^{\nu_i}\}_{i=1}^n}\right) > 1 - \epsilon .$$

Proof. According to 6.3.2 there are numbers $c_j(q)$ such that

$$c_j(q) \geq \sup \left\{ |\alpha_j| : \left\|\sum_{j=1}^p \alpha_j t^{\lambda_j}\right\|_C \leq 1 \right\} \quad \text{for all } j$$

whenever $\Lambda = \{\lambda_j\}_{j=1}^n$ satisfies $\inf \lambda_{j+1}/\lambda_j \geq q$. (Recall that $\|T_\rho\|_C \leq 1$ whenever $\rho \in [0, 1]$.) Fix $\epsilon > 0$ and consider $\Lambda_1 = \{\mu_j\}_{j=1}^m$ with $\inf_j \mu_{j+1}/\mu_j \geq q_1$ and $\mu_1 < \ldots < \mu_m \leq a$. We find $t_0 \in \,]0, 1[$ such that

$$t^{\mu_j} - t_0^{\mu_j} \leq 1 - t_0^a \leq \frac{\epsilon}{2 \max_{j \leq m} c_j(q_1) m}$$

if $t_0 \leq t \leq 1$ and $j = 1, \ldots, m$, where t_0 depends only on ϵ, q_1, m and a but not on Λ_1. Find N with

$$t_0^N \leq \frac{\epsilon}{4 \max_{k \leq n} c_k(q_2) n}$$

where N depends only on t_0, ϵ, q_2 and n. Take any $\Lambda_2 = \{\nu_1, \ldots, \nu_n\}$ with $N < \nu_1 < \ldots < \nu_n$ and $\inf \nu_{k+1}/\nu_k \geq q_2$. Put $A = \operatorname{span}\{t^{\mu_j}\}_{j=1}^m$ and $B = \operatorname{span}\{t^{\nu_k}\}_{k=1}^n$. If $f \in A$ with $\|f\|_C = 1$ then, in view of the choice of t_0, $\sup_{t \geq t_0} |f(t)| \leq |f(t_0)| + \epsilon/2$. Hence $\sup_{t \leq t_0} |f(t)| \geq \|f\|_C - \epsilon/2$. By the choice of N, with $g \in B$ and $\|g\|_C \leq 2$ we obtain $\sup_{t \leq t_0} |g(t)| \leq \epsilon/2$ and hence

$$\|f - g\|_C \geq \sup_{t \leq t_0} |f(t) - g(t)| \geq \sup_{t \leq t_0} |f(t)| - \sup_{t \leq t_0} |g(t)| \geq 1 - \epsilon .$$

Thus $(\widehat{A, B}) > 1 - \epsilon$.   □

We also need the following classical fact from approximation theory which is due to Kharshiladze-Lozinski. For a proof see [143].

Theorem 10.2.2 *Let $L_n =span\ \{t^k\}_{k=1}^n \subset C[0,1]$. Then, for some absolute constant $c > 0$,*

$$\lambda(L_n, C) = \lambda(L_n) \geq c \log n, \quad n = 1, 2 \ldots$$

Constructing special block-arithmetic Müntz sequences we prove

Theorem 10.2.3 *There exists an infinite dimensional non-dense block-lacunary Müntz space $[M(\Lambda)]_C$ in C of L-projection type. $[M(\Lambda)]_C$ has an FDD whose summands are isometrically isomorphic to $span\{t^k\}_{k=1}^n$, $n = 1, 2 \ldots$.*

Proof. Take a positive sequence $\{\beta_k\}_{k=1}^\infty$ such that $\beta := \prod_{k=1}^\infty \beta_k > 0$, and $\beta_k < 1$, $k = 1, 2, \ldots$. Consider $\epsilon > 0$. Take N_1 such that $1/N_1 < \epsilon/2$. Denote span $\{t^{N_1}\}$ by B_1.

For $\mu_1 = N_1$, $n = 2$ and $q_2 = 2$, using Theorem 10.2.1, find N_2 such that $1/N_2 + 1/(2N_2) < \epsilon/4$ and, with $B_2 := span\{t^{N_2}, t^{2N_2}\}$, we have $(\widehat{B_1, B_2}) \geq \beta_1$.

For $\mu_1 = N_1$, $\mu_2 = N_2$, $\mu_3 = 2N_2$, $n = 3$ and $q_2 = 3/2$ apply Theorem 10.2.1 to find N_3 such that $1/N_3 + 1/(2N_3) + 1/(3N_3) < \epsilon/8$ and, with $B_3 := $ span $\{t^{N_3}, t^{2N_3}, t^{3N_3}\}$, we have $(\widehat{B_1 + B_2}, B_3) \geq \beta_2$.

Continuing this process we get the sequence $\{N_k\}_{k=1}^\infty$ and the spaces $\{B_k\}_{k=1}^\infty$ with dim $B_k = k$, $k = 1, 2, \ldots$, where $B_k = $ span $\{t^{N_k}, t^{2N_k}, \ldots, t^{kN_k}\}$ such that

$$\frac{1}{N_k} + \ldots + \frac{1}{kN_k} \leq \frac{\epsilon}{2^k} \quad \text{and} \quad (\widehat{B_1 + B_2 + \ldots + B_k}, B_{k+1}) \geq \beta_k \ ,$$

for all k. Put

$$\Lambda = \cup_{k=1}^\infty \{N_k, 2N_k, \ldots, kN_k\} \ .$$

Since $\sum_{k=1}^\infty \epsilon/2^k = \epsilon$ the Müntz space $B := [M(\Lambda)]_C = \overline{B_1 + B_2 + \ldots}$ is non-dense and it has L-projection type. Indeed, using simple substitution we see that B_n is isometrically isomorphic to $span\{t^k\}_{k=1}^n$ and, by 10.2.2, $\lambda(B_n) \to \infty$ as $n \to \infty$. At the same time we have, for any k,

$$(B_1 + \ldots + B_k, \widehat{B_{k+1} + B_{k+2} + \ldots}) \geq$$

$$(B_1 + \ldots \widehat{+ B_k}, B_{k+1}) \cdot (B_1 + \ldots + \widehat{B_k + B_{k+1}}, B_{k+2}) \cdot \ldots \geq \beta.$$

Proposition 2.6.1 implies that the B_k are the summands of an FDD of B. So we obtain $\sup_k \lambda(B_k, B) \leq \infty$. Thus the projection function of B is bounded which means that B has L-projection type. $\square$

The projection type of a Banach space is preserved if we go over to an isomorphic Banach space (see 1.5.6). So the Müntz space $[M(\Lambda)]_C$ of 10.2.3 cannot be isomorphic to c_0. In particular we have, using 10.1.1,

Corollary 10.2.4 *There are at least two non-isomorphic infinite dimensional sparse Müntz spaces in $C[0,1]$.*

Conjecture. There is a continuum of pairwise non-isomorphic Müntz spaces in C.

For the isometric analogue of the conjecture see Sect. 10.4. It is known that there is a continuum of pairwise non-isomorphic subspaces in c_0. Indeed, H.P.Rosenthal communicated to us the following example.

Example. Fix $1 \leq p < 2$ and put

$$E_p = \left(\sum_{n=1}^{\infty} \oplus l_p^n \right)_{(c_0)}.$$

In view of 4.1.2, for any $\epsilon > 0$, for any n there is a subspace $F_n \subset c_0$ such that the Banach-Mazur distance satisfies $d(l_p^n, F_n) \leq 1 + \epsilon$. Hence E_p is $(1 + \epsilon)$-isomorphic to $(F_1 \oplus F_2 \oplus \ldots)_{(c_0)}$ which is isometric to a subspace of c_0. We claim that E_p and E_q are not isomorphic whenever $p \neq q$ and $2 \leq p, q < \infty$. Hence c_0 has a continuum of pairwise non-isomorphic subspaces.

To prove the claim let $2 \leq p < q < \infty$ and assume that $T : E_p \to E_q$ is an isomorphism. Fix n and let $\{e_k\}_{k=1}^n$ be the unit vector basis of the n'th component of E_p, l_p^n. Let $\{e_{j,l}\}_{l=1}^j$ be the unit vector basis of the j'th component of E_q, l_q^j. Then there are numbers $\beta_{k,j,l}$ with $Te_k = \sum_{j=1}^{\infty} \sum_{l=1}^j \beta_{k,j,l} e_{j,l}$. Hence

$$n^{1/p} = \left\| \sum_{k=1}^n \theta_k e_k \right\|_{E_p} \leq \|T^{-1}\| \sup_j \left(\sum_{l=1}^j \left| \sum_{k=1}^n \theta_k \beta_{k,j,l} \right|^q \right)^{1/q}$$

for any $\theta_k \in \{1, -1\}$. The Khintchine inequality [84] yields a universal constant $c > 0$ independent of $\beta_{k,j,l}$ and n with

$$\sum_{\theta_k \in \{1,-1\}} \frac{1}{2^n} \left| \sum_{k=1}^n \theta_k \beta_{k,j,l} \right| \leq c \left(\sum_{k=1}^n |\beta_{k,j,l}|^2 \right)^{1/2}.$$

This implies, since $q > 2$,

$$n^{1/p} \leq \|T^{-1}\| c \sup_j \left(\sum_{l=1}^j \left(\sum_{k=1}^n |\beta_{k,j,l}|^2 \right)^{q/2} \right)^{1/q}$$

$$\leq \|T^{-1}\| c \sup_j \left(\sum_{l=1}^j \sum_{k=1}^n |\beta_{k,j,l}|^q \right)^{1/q}$$

$$\leq \|T^{-1}\| \cdot \|T\| c n^{1/q}.$$

This contradicts $1/q < 1/p$ for large enough n. Hence such an ismorphism T does not exist. $\qquad \square$

10.3 Bases in Some Müntz Spaces of L-Projection Type

Here we show that the Müntz space of 10.2.3 has a basis. To this end we use a classical result of Bočkarev ([10], see also [8]). Let

$$E_n = \text{span}\{\exp(i2\pi k\varphi) \ : \ k = 0, 1, \dots, 2n\}$$

be regarded as subspace of the complex Banach space $C_{\mathbf{C}}[0,1]$. Then Bočkarev showed

Theorem 10.3.1 *[10] There are real numbers $a_{m,l}$, $m, l = 0, 1, \dots, 2n$, such that*

$$g_m(\varphi) = \sum_{l=0}^{2n} a_{m,l} \sum_{k=0}^{2n} \exp\left(ik\left(2\pi\varphi - \frac{\pi l}{n}\right)\right), \quad \varphi \in [0,1],$$

are the elements of a basis of E_n whose basis constant does not depend on n.

We need a real version of 10.3.1. Let

$$A_n = \text{span}\{\sin(2\pi k\varphi) \ : \ k = 1, \dots, n\} \subset C[0,1],$$
$$B_n = \text{span}\{\cos(2\pi k\varphi) \ : \ k = 0, 1, \dots, n\} \subset C[0,1]$$

and

$$C_n = A_n + B_n$$

Then we obtain

Corollary 10.3.2 *The spaces A_n, B_n and C_n, $n = 1, 2, \dots$, have bases with uniformly bounded basis constants.*

Proof. Let D_n be the complex version of C_n, i.e.

$$D_n = \text{span}\{\exp(i2\pi j\varphi) \ : \ j = -n, \dots, -1, 0, 1, \dots, n\} \ .$$

Of course, D_n and E_n are isometrically isomorphic. With the notation of 10.3.1 we define

$$h_m(\varphi) := \exp(-in2\pi\varphi)g_m(\varphi) = \sum_{l=0}^{2n} a_{m,l}(-1)^l \sum_{j=-n}^{n} \exp\left(ij\left(2\pi\varphi - \frac{\pi l}{n}\right)\right) \ .$$

Then $\Omega := \{h_0, h_1, \dots, h_{2n}\}$ is a basis of D_n and the h_m are real-valued. Hence Ω is as well a basis of C_n whose basis constant does not exceed the preceding one. Put, for $f \in C_n$, $(If)(\varphi) = f(1 - \varphi)$. Then $I|_{A_n} = -id$, $I|_{B_n} = id$ and we see that $(id - I)C_n = A_n$, $(I + id)C_n = B_n$. If $f \in B_n$ then there are unique $\alpha_m \in \mathbf{R}$ with

$$f = \sum_{m=0}^{2n} \alpha_m h_m = \sum_{m=0}^{2n} \alpha_m I h_m = If \ .$$

Hence $f = \sum_{m=0}^{2n} \alpha_m 2^{-1}(h_m + Ih_m)$. Clearly, for each $M < 2n$, we have $\|\sum_{m=0}^{M} \alpha_m 2^{-1}(h_m + Ih_m)\|_C \le c\|f\|_C$ where c is the basis constant of Ω. This implies that we can choose a suitable subset of $2^{-1}(id + I)(\Omega)$ to obtain a basis of B_n with basis constant smaller than or equal to c. Similarly, $2^{-1}(id - I)(\Omega)$ contains a basis of A_n whose basis constant is at most equal to c. □

Proposition 10.3.3 *Let* $L_n = span\{t^k\}_{k=1}^{n} \subset C[0,1]$. *Then we have* $\sup_n d(L_n, B_{n-1}) < \infty$ *where* $d(\cdot, \cdot)$ *is the Banach-Mazur distance. In particular, the* L_n *have bases with uniformly bounded basis constants.*

Proof. Put $\tilde{L}_n = \text{span}\{1\} + L_{n-1}$ and let $P_n : \tilde{L}_n \to \text{span}\{1\}$ be the projection with $(P_n f)(t) = f(0)$. Then $\|P_n\| = 1$ and $P_n|_{L_{n-1}} = 0$. We have $L_n = L_{n-1} + \text{span}\{t^n\}$. Let $x + L_{n-1} \in L_n/L_{n-1}$ be of norm one and find $x^* \in (L_n/L_{n-1})^* = L_{n-1}^{\perp}$ of norm one with $x^*(x) = 1$. Since $\dim L_n < \infty$ we can assume that $\|x\| = 1$. Put, for $g \in L_n$, $Q_n g = x^*(g)x$. Then Q_n is a projection with $\|Q_n\| = 1$ and $Q_n|_{L_{n-1}} = 0$. Let $S : \tilde{L}_n \to L_n$ be defined by

$$Sf = (id - P_n)f + f(0)x$$

Then $\|S\| \cdot \|S^{-1}\| \le 9$ and thus $\sup_n d(\tilde{L}_n, L_n) \le 9$.

Now it suffices to show that $\tilde{L}_{n+1}$ is isometrically isomorphic to B_n. To this end let $T_1 : \tilde{L}_{n+1} \to C[-1,1]$ be defined by

$$(T_1 f)(s) = f\left(\frac{1}{2}s + \frac{1}{2}\right), \quad s \in [-1,1]$$

Furthermore, let $T_2 : T_1\tilde{L}_{n+1} \to C[0,1]$ be the operator with

$$(T_2 g)(\varphi) = g(\cos(2\pi\varphi)), \quad g \in T_1\tilde{L}_{n+1}.$$

It is easily seen by induction that $T_2 T_1 \tilde{L}_{n+1} \subset C_n$. Moreover, by definition we have $(T_2 g)(1 - \varphi) = (T_2 g)(\varphi)$. Hence $T_2 T_1 \tilde{L}_{n+1} \subset B_n$. Since $T_2 T_1$ is an isometry and $\dim B_n = \dim \tilde{L}_{n+1}$ we obtain $T_2 T_1 \tilde{L}_{n+1} = B_n$. □

As a consequence of 10.3.2, 10.3.3 and 2.6.2 we have

Corollary 10.3.4 *Let* B *be a Banach space with an FDD whose summands are isometrically isomorphic to* L_n, $n = 1, 2, \ldots$. *Then* B *has a basis.*

Problems 10.3.5 *Let* $0 < \lambda_1 < \lambda_2 < \ldots$ *be such that* $\sum_{k=1}^{\infty} 1/\lambda_k < \infty$. *Put* $L_n(\Lambda) = span\{t^{\lambda_k}\}_{k=1}^{n} \subset C[0,1]$. *Do the* $L_n(\Lambda)$, $n = 1, 2, \ldots$, *have bases with uniformly bounded basis constants? Give estimates for* $\lambda(L_n(\Lambda), [M(\Lambda)]_C)$.

10.4 Non-Isometric Müntz Spaces in C

Here we give a positive answer to the isometric analogue of the conjecture at the end of Sect. 10.2. Namely we show that there is a continuum of non-isometric Müntz spaces in C.

At first, we consider λ_0 and λ_1 with $0 \le \lambda_0 < \lambda_1$.

Proposition 10.4.1 *If $\lambda_0 = 0$ then $e_0(t) = t^{\lambda_1}$ and $e_1(t) = 1 - t^{\lambda_1}$ are isometrically equivalent to the unit vector basis of l_∞^2. Hence $[M(\Lambda)]_C$ is isometrically isomorphic to l_∞^2.*

Proof. We verify directly that

$$||\alpha_0 e_0 + \alpha_1 e_1||_C = \max\{|\alpha_0|, |\alpha_1|\} \ .$$

for all α_0 and α_1. □

Definition 10.4.2 *The Banach space X is called* strictly normed *if for any two non-zero elements x and y in X, with $x \ne \lambda y$ for any $\lambda > 0$, we have*

$$||x + y|| < ||x|| + ||y|| \ .$$

Let us recall (see 3.3.2) that X is polyhedral if the Minkowski curve of any two-dimensional subspace is a polygon. Of course, if X is strictly normed then X is non-polyhedral. The converse is false.

Proposition 10.4.3 *Let $\Lambda = \{\lambda_0, \lambda_1\}$ be such that $0 < \lambda_0 < \lambda_1$. Then $[M(\Lambda)]_C$ is non-polyhedral. But $[M(\Lambda)]_C$ is not strictly normed.*

Proof. By substituting $s = t^{\lambda_0}$ we can reduce the proof to the case where $\Lambda = \{1, 1 + \delta\}$ for some $\delta > 0$. An elementary calculation shows that

$$\gamma(a) := ||at - t^{1+\delta}||_C = \max\left(|1 - a|, |a|^{(1+\delta)/\delta} \frac{\delta}{(1 + \delta)^{(1+\delta)/\delta}}\right)$$

for any $a \in [0, 1+\delta]$. If a is small enough we obtain that $||at - t^{1+\delta}||_C = 1 - a$. Hence $\gamma(a)$ is affine on $]0, a_0[$ for sufficiently small $a_0 > 0$ which implies that $[M(\Lambda)]_C$ is not strictly normed. For a close to one we have $\gamma(a) = |a|^{(1+\delta)/\delta} \delta(1 + \delta)^{-1-1/\delta}$ which is non-affine. We conclude that $[M(\Lambda)]_C$ is non-polyhedral. □

Propositions 10.4.1 and 10.4.3 imply

Theorem 10.4.4 *Let $\Lambda = \{\lambda_0, \lambda_1, \ldots\}$ be finite or infinite with $0 \le \lambda_0 \le \lambda_1 \le \ldots$. Then $[M(\Lambda)]_C$ is polyhedral if and only if $[M(\Lambda)]_C$ is two-dimensional and $\lambda_0 = 0$. $[M(\Lambda)]_C$ is never strictly normed.*

Now we come to the main result of this section. Let X be a Banach space and, again, for a finite dimensional subspace E of X let $\lambda(E, X)$ and $\lambda(E)$ be the relative and absolute projection constant of E. In connection with the notion of load (see 1.5.5) we shall use the following quantity

$$l_n(X) = \inf \left\{ \frac{\lambda(E, X)}{\lambda(E)} \ : \ E \text{ a subspace of } X, \ \dim E = n \right\} .$$

Theorem 10.4.5 *Let $n \geq 2$ be a finite integer or infinite. Then there is a continuum of pairwise non-isometric n-dimensional Müntz spaces in C.*

Proof. We prove the case $n = \infty$. The case of finite n is almost identical. Let $B_0 = \mathrm{span}\{1, t\}$ and $B_1 = \mathrm{span}\{t^2, t^3\}$. According to 10.4.1, B_0 is isometric to l_∞^2 and, by 10.4.3, B_1 is not isometric to l_∞^2. Hence we have $\lambda(B_0) = 1$ and $\lambda(B_1) > 1$ (see 1.5). Fix $\epsilon > 0$ such that

$$\frac{1}{(1 - \epsilon)\lambda(B_1)} < (1 - \epsilon)^2 .$$

Use the theorem on almost orthogonality, 10.2.1, and induction to find a sequence of real numbers $3 < \mu_3 < \mu_4 < \dots$ such that

$$(B_0 + \ \mathrm{span}\{t^{\mu_3}, \dots, t^{\mu_m}\}, \ \widehat{\mathrm{span}\{t^{\mu_{m+1}}, t^{\mu_{m+2}}, \dots\}}) > 1 - \epsilon ,$$

for every integer $m \geq 3$, and $(B_1, \ \widehat{\mathrm{span}\{t^{\mu_k}\}_{k=3}^\infty}) > 1 - \epsilon$. (This argument was used already in 10.2.2.) Put

$$X_0 = B_0 + \overline{\mathrm{span}}\{t^{\mu_k}\}_{k=3}^\infty \text{ and } X_1 = B_1 + \overline{\mathrm{span}}\{t^{\mu_k}\}_{k=3}^\infty .$$

Let P_m be the projection with

$$P_m \left(\alpha_0 + \alpha_1 t + \sum_{k=3}^\infty \alpha_k t^{\mu_k} \right) = \alpha_0 + \alpha_1 t + \sum_{k=3}^m \alpha_k t^{\mu_k} \quad \text{for all } \alpha_k$$

where only finitely many α_k are different from zero. By the choice of the μ_k we have $\|P_m\| \leq (1 - \epsilon)^{-1}$ for all m including $m = 0$ and $m = 1$. Since

$$\alpha_0 + \alpha_1 t + \sum_{k=3}^m \alpha_k t^{\mu_k} = \left(\sum_{k=0}^m \alpha_k \right) t^{\mu_m} + \left(\sum_{k=0}^{m-1} \alpha_k \right) (t^{\mu_{m-1}} - t^{\mu_m}) + \dots + \alpha_0 (1 - t)$$

for all m we obtain

$$(1 - \epsilon) \sup_m \left| \sum_{k=0}^m \alpha_k \right| \leq \left\| \alpha_0 + \alpha_1 t + \sum_{k=3}^\infty \alpha_k t^{\mu_k} \right\|_C \leq \sup_m \left| \sum_{k=0}^m \alpha_k \right| .$$

This shows that $d(X_0, c) \leq (1 - \epsilon)^{-1}$ where $d(\cdot, \cdot)$ is the Banach-Mazur distance. Since $l_2(c) = 1$ we obtain $l_2(X_0) \geq (1 - \epsilon)^2$ (see 1.5.6). Moreover, since $(B_1, \ \widehat{\mathrm{span}\{t^{\mu_k}\}_{k=3}^\infty}) > 1 - \epsilon$ we have $\lambda(B_1, X_1) \leq (1 - \epsilon)^{-1}$ and so

$$l_2(X_1) \leq \frac{\lambda(B_1, X_1)}{\lambda(B_1)} \leq \frac{1}{(1-\epsilon)\lambda(B_1)} < (1-\epsilon)^2 \leq l_2(X_0) .$$

This proves that X_0 and X_1 are non-isometric. Finally, for $\alpha \in [0,1]$ put $B_\alpha = \text{span}\{t^{2\alpha}, t^{1+2\alpha}\}$ and $X_\alpha = B_\alpha + \text{span}\{t^{\mu_k}\}_{k=3}^\infty$. Then $d(B_\alpha, B_0)$ and $d(X_\alpha, X_0)$ are continuous with respect to α. This implies that $l_2(X_\alpha)$ is continuous with respect to α. Hence $l_2(X_\alpha)$ attains all values in $[l_2(X_1), l_2(X_0)]$. So there is an interval $[a,b] \subset [0,1]$ such that, for each $\alpha_1, \alpha_2 \in [a,b]$ with $\alpha_1 \neq \alpha_2$, X_{α_1} and X_{α_2} are non-isometric.   □

The Classes $[\mathcal{M}]$, $\mathcal{A}$, $\mathcal{P}$ and $\mathcal{P}_\epsilon$

In this chapter we study more general classes of subspaces of $C[0,1]$ which contain the class of Müntz spaces and we discuss theorems on compactness (11.1.2 and 11.2.6) as well as on interpolation (11.3.1). Moreover, we investigate which of these spaces can be embedded into c_0. Finally we deal with notions of universality for Müntz sequences and spaces.

11.1 The Classes $[\mathcal{M}]$ and $\mathcal{A}$

We denote by $\mathcal{M}$ the class of all non-dense Müntz sequences and by $[\mathcal{M}]$ the class of all non-dense Müntz spaces in C. Let $\mathcal{M}^{gap}$ be the class of all non-dense Müntz sequences $M(\Lambda)$ where $\Lambda = \{\lambda_k\}_{k=1}^\infty$ satisfies the gap condition $\inf_k(\lambda_{k+1} - \lambda_k) > 0$ and let $[\mathcal{M}^{gap}]$ be the class of all corresponding Müntz spaces in C. Similarly, the class of sparse rational and integer Müntz sequences will be denoted by $\mathcal{M}^{rat}$ and $\mathcal{M}^{int}$, resp., and the corresponding classes of Müntz spaces in C by $[\mathcal{M}^{rat}]$ and $[\mathcal{M}^{int}]$. We also shall consider finite sequences and finite dimensional spaces which will be denoted by $\mathcal{M}^{fin}$ and $[\mathcal{M}^{fin}]$.

Sometimes it is convenient to study more general classes of subspaces of $C[0,1]$ than $[\mathcal{M}]$. Closely related to class $[\mathcal{M}]$ are the following classes. Let $D = \{z \in \mathbf{C} : |z| < 1\}$.

Definition 11.1.1 *A closed subspace X in $C[0,1]$ is said to be a* subspace of class $\mathcal{A}$ *if each function $x(t) \in X$ has an analytical extension to*

$$D \setminus \{z \in \mathbf{C} : Rez \leq 0\}$$

A closed subspace X in $C[0,1]$ is said to be a subspace of class $\mathcal{A}_D$ *if each function $x(t) \in X$ has an analytical extension to D.*

With 6.2.3 we obtain

$$[\mathcal{M}^{int}] \subset \mathcal{A}_D \subset \mathcal{A} \quad \text{and} \quad [\mathcal{M}^{int}] \subset [\mathcal{M}^{gap}] \subset \mathcal{A} .$$

For the spaces of class $\mathcal{A}$ the following compactness theorem is valid which is an anlogue of the Montel theorem for normal families of analytical functions.

Theorem 11.1.2 *a) Let $X \in \mathcal{A}$ and let $\mathcal{K} \subset X$ be a uniformly bounded family of functions in X. Then there is a subsequence $\{f_n\}_{n=1}^\infty \subset \mathcal{K}$ which converges uniformly on each interval $[a, b] \subset]0, 1[$. In other words, $\mathcal{K}|_{[a,b]}$ is relatively compact with respect to the sup-norm on $[a, b]$ for each $[a, b] \subset]0, 1[$.*
b) Let $X \in \mathcal{A}_D$ and let $\mathcal{K} \subset X$ be a uniformly bounded family of functions in X. Then there is a subsequence $\{f_n\}_{n=1}^\infty \subset \mathcal{K}$ which converges uniformly on each interval $[0, b] \subset [0, 1[$.

In 11.2.6 we shall give a proof of this theorem for a wider class of subspaces.

11.2 The Class $\mathcal{P}$

Before the introduction of $\mathcal{P}$ we need some preliminaries. We state and prove two propositions which are convenient for applications.

Proposition 11.2.1 *Let $\{f_k\}_{k=1}^\infty$ be a uniformly bounded sequence of functions in $C[a, b]$ which is not equicontinuous. Then there is a sequence $\{a_k\}_{k=1}^\infty$ with $\sum_k |a_k| < \infty$ such that the function $f(t) = \sum_{k=1}^\infty a_k f_k(t)$ is non-differentiable in some point of $[a, b]$.*

Proof. Here we find $\epsilon > 0$ and sequences $\{n_k\}_{k=1}^\infty$ and $\{(t_k, s_k)\}_{k=1}^\infty$ where $t_k, s_k \in [a, b]$ such that $t_0 = \lim_{k \to \infty} t_k = \lim_{k \to \infty} s_k$ exists but

$$|(f_{n_k}(t_k) - f_{n_k}(t_0)) + (f_{n_k}(t_0) - f_{n_k}(s_k))| \geq 2\epsilon \quad \text{for all } k .$$

By going over to a suitable subsequence we may assume without loss of generality that

$$|f_{n_j}(t_k) - f_{n_j}(t_0)| \leq \frac{\epsilon}{2^k}, \quad j = 1, \ldots, k-1, \quad |f_{n_k}(t_k) - f_{n_k}(t_0)| \geq \epsilon$$

and $|t_k - t_0| \leq 1/4^k$ for all k. Furthermore, by induction, take $\theta_k \in \{-1, 1\}$ such that

$$\sum_{j=1}^{k-1} \frac{\theta_j}{2^j}(f_{n_j}(t_k) - f_{n_j}(t_0)) \quad \text{and} \quad \frac{\theta_k}{2^k}(f_{n_k}(t_k) - f_{n_k}(t_0))$$

have the same sign. Finally put $f = \sum_{j=1}^\infty \theta_j/2^j f_{n_j}$. Then, for any $k \geq 3$, we have

$$|f(t_k) - f(t_0)| \geq \frac{1}{2^k}|f_{n_k}(t_k) - f_{n_k}(t_0)| - \sum_{j=k+1}^{\infty} \frac{1}{2^j}|f_{n_j}(t_{n_k}) - f_{n_j}(t_0)|$$

$$\geq \frac{1}{2^k}\epsilon - \sum_{j=k+1}^{\infty} \frac{1}{2^{j+k}}\epsilon$$

$$\geq \frac{1}{2^{k+1}}\epsilon.$$

This implies that $\lim_{k\to\infty}(f(t_k) - f(t_0))/(t_k - t_0)$ does not exist. □

Definition 11.2.2 *A sequence of subsets $A_k \subset [a,b]$ is called* condensed *on $[a,b]$ if, for some positive sequence $\{\epsilon_k\}_{k=1}^{\infty}$ with $\lim_{k\to\infty}\epsilon_k = 0$, each set A_k forms an ϵ_k-net on $[a,b]$, i.e. if for each $t \in [a,b]$ there is an $s \in A_k$ with $|t - s| < \epsilon_k$.*

The sequence of functions $\{f_k\}_{k=1}^{\infty}$ on $[a,b]$ will be called condensed *on $[a,b]$ if the sequence $\{N_{f_k}\}_{k=1}^{\infty}$ is condensed on $[a,b]$ where N_{f_k} is the set of all zeros of f_k in $[a,b]$.*

With these notions and Proposition 11.2.1 we have

Proposition 11.2.3 *([48], Theorem 1) Let $f_k \in C[a,b]$ be such that $\|f_k\| = 1$ for all k. If $\{f_k\}_{k=1}^{\infty}$ is condensed on $[a,b]$ then there exists a sequence of real numbers $\{a_k\}_{k=1}^{\infty}$ with $\sum_k |a_k| < \infty$ such that $f(t) = \sum_{k=1}^{\infty} a_k f_k$ is not everywhere differentiable on $[a,b]$.*

Proof. In view of Proposition 11.2.1 it suffices to asssume that $\{f_k\}_{k=1}^{\infty}$ is equicontinuous. Let $t_k \in [a,b]$ be such that $|f_k(t_k)| = 1$ for all k. Moreover, let t_0 be an accumulation point of $\{t_k\}_{k=1}^{\infty}$. Using the assumption on the condensed zero sets we find, by induction, a subsequence $\{n_k\}_{k=1}^{\infty}$ of the indices such that, for all k,

$$|t_{n_k} - t_0| \leq \frac{1}{4^k}, \quad |f_{n_k}(t_{n_j})| \leq \frac{1}{2^k}, j = 1, 2, \ldots, k-1, \quad \text{and} \quad |f_{n_k}(t_0)| \leq \frac{1}{2^k}.$$

Put $f = \sum_{j=1}^{\infty} \theta_j/2^j f_{n_j}$ where $\theta_k \in \{-1, 1\}$ is such that

$$\sum_{j=1}^{k-1} \frac{\theta_j}{2^j}(f_{n_j}(t_{n_k}) - f_{n_j}(t_0)) \quad \text{and} \quad \frac{\theta_k}{2^k}(f_{n_k}(t_{n_k}) - f_{n_k}(t_0))$$

have the same sign. Exactly as in the proof of 11.2.1 we show

$$f(t_{n_k}) - f(t_0)| \geq \frac{1}{2^{k+1}}.$$

Hence $\lim_{k\to\infty}(f(t_{n_k}) - f(t_0))/(t_{n_k} - t_0)$ does not exist. □

Since the unit ball of an infinite dimensional subspace of C is not equicontinuous we obtain

Corollary 11.2.4 *If each element $x(t)$ of a closed subspace X of C is everywhere differentiable on $[0,1]$ then dim $X < \infty$.*

For $[0,1[$ this corollary is no longer valid. Counterexamples are the sparse Müntz spaces with integer exponents (see 6.2.3).

Definition 11.2.5 *A closed subspace $X \subset C[0,1]$ will be called of class $\mathcal{P}$ (or of class $\mathcal{P}_0$, resp.) if each function $x(t)$ in X is everywhere differentiable on $]0,1[$ (or on $[0,1[$, resp.).*

$X \in \mathcal{P}$ (or $X \in \mathcal{P}_0$, resp.) will be called $\mathcal{P}$-essential, $X \in \mathcal{P}_e$ (or $\mathcal{P}_0$-essential, $X \in \mathcal{P}_{0,e}$, resp.) if, for any sequence of points t_k in $[0,1]$ with $t_1 < t_2 < \ldots < 1$, $X|_{\{t_k : k=1,2,\ldots\}}$ is infinite dimensional.

Obviously we have

$$[\mathcal{M}^{int}] \subset \mathcal{A}_D \subset \mathcal{P}_{0,e} \subset \mathcal{P} \quad \text{and} \quad [\mathcal{M}^{gap}] \subset \mathcal{A} \subset \mathcal{P}_e \subset \mathcal{P}$$

Theorem 11.2.6 *(Compactness) Every $X \in \mathcal{P}$ satisfies the assertion of Theorem 11.1.2 (a) and every $X \in \mathcal{P}_0$ satisfies the assertion of Theorem 11.1.2 (b)*

Proof. Let $\mathcal{K}$ be a uniformly bounded family of functions in X. Suppose that the restriction $\mathcal{K}|_{[a,b]}$, for some $b < 1$, is not relatively compact. This means that the family $\mathcal{K}|_{[a,b]}$ is not equicontinuous. Then, by Proposition 11.2.1, there exists a function $f \in X$ which is not differentiable in some point of $[a,b]$ and we get a contradiction. $\qquad\qquad\square$

11.3 Interpolation and Approximation in the Spaces of Class $\mathcal{P}$

From Proposition 11.2.1 we get easily

Theorem 11.3.1 *[48] Let $X \in \mathcal{P}$. Then, for any $\epsilon > 0$, $\tau \in]0, 1/2[$, $N > 0$ and any function $g \in X$ there exists $\delta = \delta(g, \epsilon, \tau, N) > 0$ satisfying the following:*

For any set of nodes in $[0,1]$ which is a δ-net for $[0,1]$ and for any $f \in X$ with $\|f\|_C \leq N$ which interpolates $g(t)$ at the given nodes we have

$$\max_{t \in [\tau, 1-\tau]} |g(t) - f(t)| < \epsilon \,.$$

Proof. By 11.2.1 the set $K := \{f|_{[\tau/2, 1-\tau/2]} \ : \ f \in X, \ \|f\|_C \leq N\}$ is equicontinuous with respect to the sup-norm on $[\tau, 1-\tau]$. Fix $\epsilon > 0$ and take $0 < \delta < \tau/2$ such that $|g(t) - g(s)| < \epsilon/2$ and $|f(t) - f(s)| < \epsilon/2$ whenever $f \in K$ and $t, s \in [\tau/2, 1-\tau/2]$, $|t-s| < \delta$. Then we obtain, for every δ-net $\Omega \subset [0,1]$, that $|f(t) - g(t)| < \epsilon$ if $t \in [\tau, 1-\tau]$ and if $f \in X$ interpolates g at the nodes in Ω and satisfies $\|f\|_C \leq N$. $\qquad\qquad\square$

This theorem allows us to split the elements of X into "almost pairwise disjoint" summands. In particular, for Müntz spaces, we recover a part of 9.1.6 (d):

Corollary 11.3.2 *Every infinite dimensional $X \in \mathcal{P}$, for every $\epsilon > 0$, contains a subspace which is $(1 + \epsilon)$-isomorphic to c_0.*

Proof. Fix $0 < \epsilon < 1$. Consider functions $f_1, \ldots, f_{n-1} \in X$ such that $\|f_k\|_C = 1$ and $f_k(0) = f_k(1) = 0$ for $k = 1, \ldots, n-1$. Moreover let τ_k, $k = 1, \ldots, n-1$, be elements of $]0, 1/2[$ with $\tau_1 > \tau_2 > \ldots > \tau_{n-1}$ such that $|f_k(t)| < \epsilon 4^{-k}$ if $t \in [\tau_k, 1 - \tau_k]$ and $|f_k(t)| < \epsilon 4^{-k}$ if $t \in [0, \tau_m] \cup [1 - \tau_m, 1]$ and $m > k$. Then fix $0 < \tau_n < \tau_{n-1}$ such that $|f_k(t)| \leq \epsilon 4^{-n}$ if $t \in [0, \tau_n] \cup [1 - \tau_n, 1]$, $k = 1, \ldots, n-1$. Apply 11.3.1 with $g = 0$, $N = 1$, $\tau = \tau_n$ and $\epsilon 4^{-n}$. Find $\delta > 0$ satisfying the assertion of Theorem 11.3.1. Fix any δ−net in $[0, 1]$ including 0 and 1. Since X is infinite dimensional we find $f_n \in X$ with $f_n \neq 0$ which interpolates 0 at the given nodes. We may clearly assume that $\|f_n\|_C = 1$. Hence f_n satisfies $f_n(0) = f_n(1) = 0$, $\|f_n\|_C = 1$ and $|f_n(t)| \leq \epsilon 4^{-n}$ if $t \in [\tau_n, 1 - \tau_n]$. So, by induction, we obtain $f_k \in X$ such that $\|f_k\|_C = 1$ and $|f_k(t)| \leq \epsilon 4^{-k}$ if $t \in [\tau_m, 1 - \tau_m] \setminus [\tau_{m-1}, 1 - \tau_{m-1}]$ and $k \neq m$. Hence

$$\left(1 - \frac{\epsilon}{3}\right) \max_k |\alpha_k| \leq \left\| \sum_{k=1}^{\infty} \alpha_k f_k \right\|_C \leq \left(1 + \frac{\epsilon}{3}\right) \max_k |\alpha_k| \quad \text{for all } \alpha_k$$

which proves that $\overline{\mathrm{span}}\{f_k\}_{k=1}^{\infty}$ is $(1 + \epsilon)$-isomorphic to c_0.   □

In the following section we shall show that $X \in \mathcal{P}$ is always isomorphic to a subspace of c_0 which implies again that X contains a $(1 + \epsilon)$-isomorphic copy of c_0 ([84], I.2.a.3. and I.2.e.3.). However the proof of 11.3.2 is more constructive.

11.4 Embeddings of the Spaces $X \in \mathcal{P}$ into c_0

The following theorem and its proof was communicated to us by P. Wojtaszczyk.

Theorem 11.4.1 *If $X \in \mathcal{P}$ then X is isomorphic to a subspace of c_0.*

Proof. Fix $\epsilon > 0$. To prove the theorem it suffices to consider

$$f \in X_0 = \{g \in X \; : \; g(1) = 0\}$$

since X_0 is at most one-codimensional in X. For such f let $f_n = f|_{I_n}$, $n = 1, 2, \ldots$, where

$$I_n = \left[1 - \frac{1}{n}, 1 - \frac{1}{n+1}\right].$$

According to 11.2.6 the set $\{f_n \; : \; f \in X, \|f\|_C \leq 1\}$ is a relatively compact set in $C_n := C(I_n)$. Let $\| \cdot \|_n$ be the sup-norm on I_n. Then there exists a projection $P_n : C_n \to C_n$ with $P_n(C_n) \cong l_{\infty}^{k_n}$, $\|P_n\|_C = 1$ and

$$||P_n f_n - f_n||_n \leq \epsilon ||f||_C \quad \text{for all} \quad f \in X_0 .$$

(Take functions $e_j \in C(I_n)$ with suitably small supports and $t_j \in I_n$ with

$$e_j(t_j) = 1 = ||e_j||_n, \quad \sum_{j=1}^{k_n} |e_j| = 1, \quad e_j(t_k) = 0 \quad \text{if} \quad k \neq j$$

and put $P_n g = \sum_{j=1}^{k_n} g(t_j) e_j$.)

Consider the map

$$T : X_0 \rightarrow \left(\sum_n \oplus l_\infty^{k_n} \right)_{(c_0)} \cong c_0$$

defined by $Tf = \{P_n f_n\}_{n=1}^\infty$, $f \in X_0$. Clearly, $\lim_n ||P_n f_n||_n = 0$ since $\lim_n ||f_n||_n = 0$,

$$||Tf||_C = \sup_n ||P_n f_n||_n \leq \sup_n ||f_n||_n = ||f||_C$$

and

$$||Tf||_C = \sup_n ||P_n f_n||_n \geq \sup_n (||f_n||_n - \epsilon ||f||_C) \geq (1 - \epsilon) ||f||_C .$$

This proves the theorem.   □

The proof shows that it is even possible to construct a $(1 + \epsilon)$-isomorphism from X onto a suitable subspace of c_0 for any $\epsilon > 0$. The assertion of this stronger version of Theorem 11.4.1 was first suggested to us by Dirk Werner (Berlin).

In the preceding chapters we also dealt with Müntz spaces in L_p. One can define analogues of the classes $\mathcal{A}$ and $\mathcal{P}$ for subspaces of L_p. It is plausible that one gets results similar 11.4.1 for embeddings in l_p.

11.5 Universality for Finite Müntz Sequences

Now we deal with $\mathcal{M}^{fin}$ and $[\mathcal{M}^{fin}]$. This section is related to questions about the existence of universal or almost universal sparse Müntz sequences (or spaces) for the class of all finite Müntz sequences in C (or finite dimensional Müntz spaces).

At first, in accordance with the notion of a universal Banach space (Definition 4.1.1), using Definition 2.7.1, we introduce the notion of universal sequence in a Banach space.

Definition 11.5.1 *Let $\bar{u} = \{u_k\}_{k=1}^{\infty}$ be a normalized sequence in a Banach space and let $\mathcal{B}$ be a set of sequences in some Banach spaces. We call $\bar{u}$ a-universal for $\mathcal{B}$ if for any $\bar{v} \in \mathcal{B}$ there is a subsequence of $\bar{u}$ which is a-equivalent to $\bar{v}$. If $\bar{u}$ is a-universal for $\mathcal{B}$ with respect to every $a > 1$ then $\bar{u}$ is called* almost universal *for $\mathcal{B}$. Finally, if $\bar{u}$ is 1-universal for $\mathcal{B}$ we simply say that $\bar{u}$ is* universal *for $\mathcal{B}$.*

For example, 4.4.1 shows that there are bases which are almost universal for the class of all monotone bases in Banach spaces.

For finite rational Müntz sequences we have

Theorem 11.5.2 *There exists a sparse integer sequence $U = \{u_k\}_{k=1}^{\infty}$ such that each finite rational Müntz sequence $\bar{t} = \{t^{r_k}\}_{k=1}^{n}$, where $r_1 < \ldots < r_n$, is isometrically equivalent to a subsequence $M(\{u_{j_k}\}_{k=1}^{n})$ of $M(U)$ in C. The space $[M(U)]_C$ is universal for the class of all rational finite dimensional Müntz spaces in C.*

Proof. Let $\rho_1 = m_1/n_1, \rho_2 = m_2/n_2, \ldots$, be the sequence of all positive rational numbers, where m_k and n_k are positive integers. Let $R_1, R_2, \ldots$ be the sequence of all increasing finite rational sequences, say

$$R_k = \left\{ \frac{m_{k,1}}{n_{k,1}}, \frac{m_{k,2}}{n_{k,2}}, \ldots, \frac{m_{k,l_k}}{n_{k,l_k}} \right\}, k = 1, 2, \ldots .$$

Let $\{b_k\}_{k=1}^{\infty}$ and $\{N_k\}_{k=1}^{\infty}$ be increasing sequences of positive integers such that

$$b_k \leq \min_{1 \leq j \leq l_k} N_k \frac{m_{k,j}}{n_{k,j}} \leq \max_{1 \leq j \leq l_k} N_k \frac{m_{k,j}}{n_{k,j}} < b_{k+1} \text{ for all } k \text{ and } \sum_{k=1}^{\infty} \frac{l_k}{b_k} < \infty$$

and such that N_k is a common denominator of all fractions in R_k. Put

$$u_k^{(j)} = N_k \frac{m_{k,j}}{n_{k,j}}, \ j = 1, \ldots, l_k, \ k = 1, 2, \ldots .$$

Finally, let $U = \{u_k\}_{k=1}^{\infty}$ be the result of a renumeration of the $u_k^{(j)}$ as follows

$$u_1^{(1)}, \ldots, u_1^{(l_1)}, u_2^{(1)}, \ldots, u_2^{(l_2)}, \ldots$$

The condition on the b_k ensures that U is sparse. It follows from the construction, with Proposition 7.3.2, that U is a universal sequence of integers. □

The preceding theorem implies

Corollary 11.5.3 *Let U be the universal sequence of Theorem 11.5.2. Then $M(U)$ is almost universal for all finite Müntz sequences in C and $[M(U)]_C$ is almost universal for the class of all finite dimensional Müntz spaces.*

Problems 11.5.4 *Does there exist a sparse Müntz sequence in C (or Müntz space) which is a-universal for all infinite sparse Müntz sequences in C (or infinite dimensional Müntz spaces)?*

Since c_0 is almost universal for the class of all finite dimensional Banach spaces 11.3.2 implies that every $X \in \mathcal{P}$, hence every infinite dimensional sparse Müntz space (satisfying the gap condition), is almost universal for the class of all finite dimensional Müntz spaces.

11.6 Sequences of Universal Disposition

Presently it is unknown if there are analogues of 11.5.2 and 11.5.3 with respect to some kind of universal disposition for the class of finite Müntz sequences or finite dimensional spaces in the spirit of Chap. 4.

Let $\bar{u} = \{u_k\}_{k=1}^\infty$ be a normalized sequence in a Banach space and $\mathcal{B}$ a set of finite sequences in some Banach spaces. As an analogue of 4.1.3 we introduce

Definition 11.6.1 *Let $a \geq 1$. We call $\bar{u}$ a sequence of a-universal disposition with respect to $\mathcal{B}$ if for each sequence $\bar{v} = \{v_i\}_{i=1}^n \in \mathcal{B}$, any subset $\{p_1, \dots, p_m\} \subset \{1, \dots, n\}$ and any isomorphism $T : span\ \{v_{p_i}\}_{i=1}^m \to span\ \bar{u}$ with $Tv_{p_i} = u_{q_i}$ for some q_i, $i = 1, \dots, m$, there exists an extension to an isomorphism $\tilde{T} : span\ \bar{v} \to span\ \bar{u}$ such that $Tv_j = u_{k_j}$ for some k_j, $j = 1, \dots, n$ and*

$$\|\tilde{T}\| \leq a\|T\|, \quad \|\tilde{T}^{-1}\| \leq a\|T^{-1}\| .$$

$\bar{u}$ is called of almost universal disposition (or of universal disposition, resp.) if $\bar{u}$ is of a-universal disposition for all $a > 1$ (or for $a = 1$, resp.).

Conjecture. There is no sparse Müntz sequence which has almost universal or a-universal disposition for some $a \geq 1$ with respect to the class of all finite Müntz sequences.

If such a Müntz sequence of almost universal disposition existed then it would be almost universal for the class of all sparse Müntz sequences.

The situation becomes clearer if we replace "sequence" by "space". A Müntz space $X \in [\mathcal{M}^{gap}]$ which is of almost universal disposition for all finite dimensional Müntz spaces in C cannot exist. Indeed, in view of 11.3.1, since c_0 is almost universal for all finite dimensional Banach spaces, X would be of almost universal disposition for all finite dimensional Banach spaces. According to 4.3.2, X would contain an isometric copy of C which would contradict 11.4.1.

12

Finite Dimensional Müntz Limiting Spaces in C

The results of this section were obtained by the first-named author and B.Shekhtman. They appear here for the first time.

It turns out that many sparse Müntz spaces share some geometric properties. Here we study the limiting behaviour of a Müntz sequence $M(\Lambda) = \{t^{\lambda_k}\}_{k=1}^{\infty}$ in C. Fix n and put $B_n^m(\Lambda) = \operatorname{span}\{t^{\lambda_{m+1}}, \ldots, t^{\lambda_{m+n}}\}$. Then we investigate the limit $B_n(\Lambda)$ of the sequence $B_n^m(\Lambda)$, $m = 1, 2, \ldots$, with respect to $\log d(\cdot, \cdot)$ where d is the Banach-Mazur distance, i.e. where

$$\lim_{m \to \infty} d(B_n(\Lambda), B_n^m(\Lambda)) = 1 \,,$$

provided such a limit exists. For surprisingly many different sequences Λ the corresponding spaces $B_n(\Lambda)$ (for fixed n) are isometric to each other.

We use the notion of limiting space from 3.3. Recall, for a sequence of n-dimensional Banach spaces B_n^m with normalized bases $\bar{e}_m = \{e_k^m\}_{k=1}^{n}$ the limiting space of $\{(B_n^m, \bar{e}_m)\}_{m=1}^{\infty}$ is $\mathbf{R}^n$ endowed with the (semi-)norm

$$\|(\alpha_1, \ldots, \alpha_n)\| = \lim_{m \to \infty} \left\| \sum_{k=1}^{n} \alpha_k e_k^m \right\|$$

provided this limit exists for all $(\alpha_1, \ldots, \alpha_n) \in \mathbf{R}^n$. The limiting space coincides with the limit with respect to the logarithm of the Banach-Mazur distance if the bases $\bar{e}_m$ are totally minimal (see 3.3.4).

12.1 Angle Convergence

At first we introduce the concept of "angle-convergence". Recall that the angle between two elements x and y of a Banach space X was defined as

$$\varphi(x, y) = \left\| \frac{x}{\|x\|} - \frac{y}{\|y\|} \right\| \qquad \text{(see 1.2)}$$

Definition 12.1.1 *We say that the sequence $\{x_m\}_{m=1}^\infty$ of non-zero elements in X angle-converges to $x \in X$ if* $\lim_{m\to\infty} \varphi(x_m, x) = 0$. *In this case we write*

$$\vee - \lim_{m\to\infty} x_m = x \quad or \quad x_m \xrightarrow{\vee} x \quad as\ m \to \infty\ .$$

More generally, let $\delta \in \mathbf{R}$ and $\delta_0 \in \mathbf{R} \cup \{\pm\infty\}$. We write $x_\delta \xrightarrow{\vee} x$ as $\delta \to \delta_0$ if $\vee - \lim_{j\to\infty} x_{\delta_j} = x$ whenever $\lim_{j\to\infty} \delta_j = \delta_0$.

The set $\{e_k\}_{k=0}^n$ of elements in X is called limiting *for the family $\{e_k^\delta\}_{k=0}^n$ if*

$$e_k^\delta \xrightarrow{\vee} e_k \quad as\quad \delta \to \delta_0 \quad for\ all\ k = 0,\ldots,n\ .$$

We note the straightforward

Lemma 12.1.2 *Let $x, x_\delta \in X \setminus \{0\}$. Then $x = \vee - \lim_{\delta\to\delta_0} x_\delta$ if and only if there are $a_\delta > 0$ and $a > 0$ with $\lim_{\delta\to\delta_0} \|ax - a_\delta x_\delta\| = 0$*

If we have angle convergence then the limiting spaces and the limits with respect to $\log d(\cdot, \cdot)$ coincide.

Proposition 12.1.3 *Let $\{e_k\}_{k=0}^n$ be a basis of the $n+1$-dimensional subspace $B_{n+1} \subset X$ and let $\{B_{n+1}^\delta\}$ be a family of $n + 1$-dimensional subspaces of X with bases $\{e_k^\delta\}_{k=0}^n$, $j = 1, 2, \ldots$, such that $e_k = \vee - \lim_{\delta\to\delta_0} e_k^\delta$ for all k. Then*

$$\Theta(B_{n+1}^\delta, B_{n+1}) \to 0 \quad and \quad d(B_{n+1}^\delta, B_{n+1}) \to 1 \quad as\ \delta \to \delta_0$$

(Here $\Theta(\cdot, \cdot)$ is the ball opening.)

Moreover B_{n+1} is the limiting space for $\delta \to \delta_0$ with respect to B_{n+1}^δ and $\{e_k^\delta/\|e_k^\delta\|\}_{k=0}^n$.

Proof. By assumption $\lim_{\delta\to\delta_0} e_k^\delta/\|e_k^\delta\| = e_k/\|e_k\|$ for every $k = 0, 1, \ldots, n$. Since $\{ e_k/\|e_k\| : k = 0, \ldots, n \}$ is a basis of B_{n+1}, for some neighbourhood U of δ_0, the bases $\bar{e}^\delta = \{ e_k^\delta/\|e_k^\delta\| : k = 0, \ldots, n \}$, $\delta \in U$, are ρ-minimal for some common $\rho > 0$ according to 2.7.2. This implies

$$\lim_{\delta\to\delta_0} \Theta(B_{n+1}^\delta, B_{n+1}) = 0 \quad and \quad \lim_{\delta\to\delta_0} d(B_{n+1}^\delta, B_{n+1})\ .$$

Moreover the bases $\bar{e}^\delta$ are totally minimal (see 3.3.4) which proves the last part of Proposition 12.1.3. □

We call the set $\{e_k\}_{k=0}^n$ of Proposition 12.1.3 a *limiting basis* for the bases $\{e_k^\delta\}_{k=0}^n$.

In the following we are concerned with $X = C[0, 1]$. Here norm convergence means uniform convergence on $[0, 1]$.

Example. Consider the finite arithmetic Müntz sequence

$$t,\ t^{1+\delta},\ t^{1+2\delta}, \ldots, t^{1+n\delta} \quad for\ some\ \delta > 0$$

which is a basis for the $n + 1$-dimensional Müntz space B_{n+1}^δ in C. In the next section we shall prove that, with respect to the logarithm of the Banach-Mazur distance, $B_{n+1} := \lim_{\delta \to 0} B_{n+1}^\delta$ exists and we shall exactly describe B_{n+1}. We cannot apply 3.3.4 directly since the bases $\{t^{1+k\delta}\}_{k=0}^n$ (depending on δ) are not totally minimal. (The functions $t^{1+k\delta}$ converge uniformly to t for any k as $\delta \to 0$. But $\{t\}$ is not a basis for any $n + 1$-dimensional space.) However we can find another family of bases for the spaces B_{n+1}^δ which will be totally minimal for sufficiently small δ.

12.2 Special Limiting Müntz Spaces of Finite Dimension

At first we need some information about finite differences.

Let $S_0 = \{f_0, f_1, f_2, \ldots\}$ be a given sequence of elements in a Banach space X. Put

$$S_1 = \{f_1 - f_0, f_2 - f_1, f_3 - f_2, \ldots\},$$

$$S_2 = \{(f_2 - f_1) - (f_1 - f_0), (f_3 - f_2) - (f_2 - f_1), \ldots\} \text{ etc.}$$

Let F_k be the initial element of the sequence S_k, for example $F_0 = f_0$, $F_1 = f_1 - f_0$ and $F_2 = (f_2 - f_1) - (f_1 - f_0) = f_0 - 2f_1 + f_2$. It is easy to prove by induction that we have

$$F_k = (-1)^k \sum_{j=0}^k (-1)^j \binom{k}{j} f_j$$

Thus we have an operator Δ (called Δ-procedure) on the set of all sequences in X which sometimes improves certain properties of the given sequences S_0. We will call $\{F_k\}_{k=0}^\infty$ the Δ-improvement of the sequence S_0. In the following we always consider the sup-norm $\| \cdot \|_C$ on $[0,1]$.

Proposition 12.2.1 *Let* $S_0 = \{t, t^{1+\delta}, t^{1+2\delta}, \ldots, t^{1+n\delta}\}$, *for some positive integer n and some $\delta > 0$. Furthermore put*

$$e_0^\delta = F_0, \ e_1^\delta = \frac{1}{\delta}F_1, \ e_2^\delta = \frac{1}{\delta^2}F_2, \ldots, e_n^\delta = \frac{1}{\delta^n}F_n \ .$$

Then the family of bases $\{e_j^\delta\}_{j=0}^n$ in $B_{n+1}^\delta = $ span $\{e_j^\delta\}_{j=0}^n$ is totally minimal for $0 < \delta < \delta_0$ and some $\delta_0 > 0$. The limiting space $B_{n+1} = \lim_{\delta \to 0} B_{n+1}^\delta$ exists and is spanned by

$$t, \ t\log t, \ t\log^2 t, \ldots, t\log^n t \ .$$

which is a limiting basis for the family $\{e_k^\delta\}_{k=0}^n$.

Proof. We have

$$F_k(t) = (-1)^k \sum_{j=0}^{k} (-1)^j \binom{k}{j} t^{1+j\cdot\delta} = (-1)^k t (1 - t^\delta)^k \sim t\delta^k \log^k t \text{ for } \delta \to 0 .$$

Hence $\lim_{\delta \to 0} e_k^\delta(t) = t \log^k(t)$ uniformly in $t \in [0,1]$ which, by 12.1.2 and 12.1.3 proves the theorem since $t, t \log t, \ldots, t \log^n t$ are linearly independent and hence form a basis in their span. $\qquad\square$

Corollary 12.2.2 *Put, for* $0 < m, n$, $L_{n+1}^m = span\{t^m, t^{m+1}, \ldots, t^{m+n}\}$ *and let* $B_{n+1} = span\{t, t \log t, \ldots, t \log^n t\}$. *Then* $\lim_{m \to \infty} d(L_{n+1}^m, B_{n+1}) = 1$.

Proof. Take the substitution $\tau = t^m$, $\delta = 1/m$ and apply 12.2.1. Note that $span\{t^m, t^m \log t, \ldots, t^m \log^n t\}$ is isometrically isomorphic to B_{n+1}. $\qquad\square$

If we go over from arithmetic to geometric Müntz sequences then we obtain the same limiting spaces but completely different limiting bases.

Theorem 12.2.3 *Let, for some integer* $n > 0$ *and some* $\delta > 0$,

$$S_0 = \{ t, t^{1+\delta}, t^{(1+\delta)^2}, \ldots, t^{(1+\delta)^n} \} .$$

Furthermore put, for the Δ-*improved family* $\{F_k\}_{k=0}^n$ *of* S_0,

$$e_0^\delta = F_0, \ e_1^\delta = \frac{1}{\delta} F_1, \ e_2^\delta = \frac{1}{\delta^2} F_2, \ldots, e_n^\delta = \frac{1}{\delta^n} F_n .$$

Then the limiting space $B_{n+1} = \lim_{\delta \to 0} span \ \{e_k^\delta\}_{k=0}^n$ *exists and again is spanned by*

$$\Omega = \{ t, \ t \log t, \ t \log^2 t, \ldots, t \log^n t \} .$$

However, the limiting basis with respect to the family $\{e_k^\delta\}_{k=0}^n$ *is different from* Ω.

Proof. Using the Taylor formula we obtain

$$t^\alpha = e^{\alpha \log t} = \sum_{l=0}^{k-1} \frac{(\alpha \log t)^l}{l!} + \frac{(\alpha \log t)^k}{k!} t^{\theta(\alpha, t)}$$

for some $\theta(\alpha, t) \in [0, \alpha]$. Hence

$$t^{(1+\delta)^j - 1} = \sum_{l=0}^{k-1} \frac{\log^l t}{l!} \left((1+\delta)^j - 1\right)^l + \frac{\log^k t}{k!} \left((1+\delta)^j - 1\right)^k t^{\theta_j}$$

$$= \sum_{l=0}^{k-1} \sum_{m=0}^{l} \binom{l}{m} (1+\delta)^{jm} (-1)^{l-m} \frac{\log^l t}{l!} + ((1+\delta)^j - 1)^k t^{\theta_j} \frac{\log^k t}{k!}$$

for some $\theta_j \in [0, (1+\delta)^j - 1]$. Thus we have, for $k \geq 1$,

$$
\begin{aligned}
e_k^\delta(t) &= \frac{(-1)^k}{\delta^k} \sum_{j=0}^{k} \binom{k}{j} (-1)^j \cdot t^{(1+\delta)^j - 1} \\
&= \frac{(-1)^k}{\delta^k} \sum_{l=0}^{k-1} \sum_{m=0}^{l} \sum_{j=0}^{k} \binom{l}{m} \binom{k}{j} (-1)^{l-m} (1+\delta)^{jm} (-1)^j \frac{t \log^l t}{l!} \\
&\quad + (-1)^k \sum_{j=0}^{k} \binom{k}{j} \left(\frac{(1+\delta)^j - 1}{\delta} \right)^k (-1)^j t^{\theta_j} \frac{t \log^k t}{k!} \\
&= \sum_{l=0}^{k-1} \sum_{m=0}^{l} \binom{l}{m} (-1)^{l-m} \left(\frac{(1+\delta)^m - 1}{\delta} \right)^k \frac{t \log^l t}{l!} \\
&\quad + (-1)^k \sum_{j=0}^{k} \binom{k}{j} \left(\frac{(1+\delta)^j - 1}{\delta} \right)^k (-1)^j t^{\theta_j} \frac{t \log^k t}{k!}
\end{aligned}
$$

Letting $\delta \to 0$ we see that e_k^δ converges uniformly to

$$
e_k(t) = \sum_{l=0}^{k-1} \sum_{m=0}^{l} \binom{l}{m} (-1)^{l-m} m^k \frac{t \log^l t}{l!} + (-1)^k \sum_{j=0}^{k} \binom{k}{j} (-1)^j j^k \frac{t \log^k t}{k!}
$$

Let $\{\tilde{e}_k\}_{k=0}^n$ be the limiting basis of 12.2.1, i.e. $\tilde{e}_k = t \log^k t$ for all k. Then we have $\tilde{e}_k \neq e_k$ for $k > 1$. On the other hand the sequences $\{\tilde{e}_k\}_{k=0}^n$ and $\{e_k\}_{k=0}^n$ are similar (see 2.1) which completes the proof of Theorem 12.2.3. $\square$

The proof shows that the limiting basis of $\{e_k^\delta\}_{k=0}^n$ in the preceding theorem consists of the functions

$$
e_k(t) := \sum_{l=0}^{k} \sum_{m=0}^{l} \binom{l}{m} (-1)^{l-m} m^k \frac{t \log^l t}{l!}
$$

Using in 12.2.3 the substitution $\tau = t^{m^n}$ and $\delta = 1/m$ we derive

Corollary 12.2.4 *Put, for* $m, n > 0$,

$$
H_{n+1}^m = \ \text{span}\{ \ t^{m^j (1+m)^{n-j}} \ : \ j = 0, 1, \ldots, n \ \}
$$

and

$$
B_{n+1} = \ \text{span}\{ \ t \log^j t \ : \ j = 0, 1, \ldots, n \ \} .
$$

Then $\lim_{m \to \infty} d(H_{n+1}^m, B_{n+1}) = 1$.

More generally we have

Theorem 12.2.5 *Let* $p(x) = \sum_{j=0}^{s} a_j x^{\mu_j}$, $x > 0$, *where* $0 < \mu_1 < \ldots < \mu_s$ *and the* a_j, $j = 0, \ldots, s$, *are given real numbers with* $a_s \neq 0$. *For fixed* $m, n > 0$ *put*

$$S_0 = \{t, t^{p(m+1)/p(m)}, \ldots, t^{p(m+n)/p(m)}\}.$$

Let $\{F_k\}_{k=0}^{n}$ *be the* Δ-*improved family of* S_0 *and put*

$$e_k^m = \mu_s^k m^k \cdot F_k, \quad k = 0, 1, \ldots, n.$$

Then the limiting space $\lim_{m\to\infty} \mathrm{span}\{e_k^m\}_{k=0}^{n}$ *exists and is spanned by*

$$\{t, t\log t, \ldots, t\log^n t\}.$$

Proof. Consider the function $\Phi(x) = t^{p(x)/p(m)}$. Using induction we obtain, for each $k \geq 1$, Müntz polynomials $q_{k,j}$ of degree less than or equal to $j\mu_s - k$ which are independent of m and satisfy

$$\frac{d^k\Phi}{dx^k} = \sum_{j=1}^{k-1} \frac{q_{k,j}(x)}{p(m)^j} t^{p(x)/p(m)} \log^j t + \left(\frac{p'(x)}{p(m)}\right)^k t^{p(x)/p(m)} \log^k t.$$

So,

$$a_{k,j} := \lim_{m\to\infty} \frac{q_{k,j}(m)}{p(m)^j} m^k \mu_s^k \quad \text{exists.}$$

An application of Taylor's formula to $\Phi(x)$ yields, for each $k \geq 1$, a number $\theta_k \in [m, m+k]$ with

$$F_k(t) = \frac{d^k\Phi}{dx^k}\big|_{x=\theta_k}$$

(See [31]). Hence

$$e_k^m(t) = \sum_{j=1}^{k-1} \frac{q_{k,j}(\theta_k)}{p(m)^j} m^k \mu_s^k t^{p(\theta_k)/p(m)} \log^j t + \left(\frac{p'(\theta_k)}{p(m)}\right)^k m^k \mu_s^k t^{p(\theta_k)/p(m)} \log^k t$$

converges to $\sum_{j=1}^{k-1} a_{k,j} t\log^j t + t\log^k t$ uniformly on $[0,1]$ as $m \to \infty$. In view of 12.1.2 and 12.1.3 this proves the theorem. $\square$

With a suitable substitution we see that 12.2.1 is a special case of 12.2.5. To illustrate the methods at a simple example we stated and proved 12.2.1 separately.

With the substitution $t = \tau^{1/p(m)}$ we obtain

Corollary 12.2.6 *Let* p *be as in 12.2.5. Put* $\Lambda = \{p(m)\}_{m=1}^{\infty}$. *Fix* n *and put*

$$B_{n+1}^m(\Lambda) = \mathrm{span}\{t^{p(m)}, \ldots, t^{p(m+n)}\}, \quad B_{n+1} = \mathrm{span}\{t, t\log t, \ldots, t\log^n t\}.$$

Then $\lim_{m\to\infty} d(B_{n+1}^m(\Lambda), B_{n+1}) = 1$.

In particular, Corollary 12.2.6 applies to μ-arithmetic sequences $\Lambda = \{m^\mu\}_{m=1}^\infty$.

The preceding proof shows that 12.2.5 remains true if $p :]0, \infty[\rightarrow]0, \infty[$ is a smooth function with

$$\lim_{x \to \infty} \frac{p(x+a)}{p(x)} = 1 \text{ for every } a > 0 \text{ and } \frac{p^{(k+1)}(x)}{p^{(k)}(x)} = O\left(\frac{p^{(k)}(x)}{p^{(k-1)}(x)}\right) \text{ as } x \to \infty$$

for every $k \geq 1$. So in particular 12.2.6 is true for $\Lambda = \{m^2 \log(m+1)\}_{m=1}^\infty$.

Problems 12.2.7 *Let*

$$L_n = span\{t^k\}_{k=1}^n, \quad L_n^m = span\{t^{m+k}\}_{k=0}^{n-1} \quad and \quad B_n = span\{t \log^k t\}_{k=0}^{n-1}.$$

Is it true that $\sup_n \sup_m d(L_n, L_n^m) < \infty$ *or* $\sup_n d(B_n, L_n) < \infty$?

The substitution $e^{-s} = t$ might be of some help here. Note that B_n is isometric to the space $span\{e^{-s}, se^{-s}, \ldots, s^n e^{-s}\} \subset C[0, \infty]$.

It might be promising to study other improving procedures, different from Δ, for example in connection with Müntz-Legendre polynomials. For finite trigonometric sequences much more is known in comparison with finite Müntz sequences, see [105].

References

1. Almira, J.M.: On Müntz theorem for countable compact sets, preprint
2. Badkov, V.M.: Convergence in the mean and almost everywhere of Fourier series in polynomials orthogonal on the interval, Math. USSR Sbornik, **24**, 223–256, (1974)
3. Banach, S.: A Course of Functional Analysis, Kiev, (1948)
4. Beauzamy, B., Lapresté, J.-T.: Modèles étalés des espaces de Banach, Hermann, Paris, (1984)
5. Bessaga, C., Pelczynski, A.: On bases and unconditional convergence of series in Banach spaces, Studia Math., **17**,151–164, (1958)
6. Bonnesen, T., Fenchel, W.: Theorie der konvexen Körper, Springer, Berlin Heidelberg New York, (1974)
7. Bourgain, J.: A remark on finite dimensional $\mathcal{P}_\lambda$-spaces, Studia Math., **72**, 285–289, (1981)
8. Bourgain, J.: Homogeneous polynomials on the ball and polynomial bases, Israel J. Math., **68**, 327–347, (1989)
9. Bočkarev, S.V.: Existence of a basis in the space of functions analytic in the disk and some properties of Franklin's system, Math. USSR Sbornik, **24**, 1–10, (1974)
10. Bočkarev, S.V.: Construction of polynomial bases in finite dimensional spaces of functions analytic in the disk, Proc. Steklov Inst. Math., **2**, 55–81, (1985)
11. Borwein, P.B., Erdélyi, T.: Lacunary Müntz sequences, Proc. Edinburgh Math. Soc. **36**, 361–374, (1993)
12. Borwein, P.B., Erdélyi, T.: Polynomials and Polynomial Inequalities, Springer, Berlin Heidelberg New York, (1995)
13. Borwein, P.B., Erdélyi, T.: Generalizations of Müntz's theorem via a Remez-type inequality for Müntz spaces, Journal of the AMS, **10**, 327–349, (1997)
14. Borwein, P.B., Shekhtman, B.: The density of rational functions: A counterexample to a conjecture of D.J.Newman, Constr. Approx., **9**, 105–110, (1993)
15. Bourgain, J., Milman, V.: Dichotomie du cotype pour les espaces invariants, C.R.Acad.Sc. Paris, **300** , 263–266, (1985)
16. Bui Min Chy, Gurariy, V.I.: Some characterizations of normed spaces and generalization of the Parseval equality on Banach spaces, Theory of Functions, Functional Analysis and Applications, **5**, 74–91, (1969)

17. Casazza, P.G., Kalton, N.J.: Notes on approximation properties in separable Banach spaces, In: P.F.X.Müller and W.Schachermeyer, editors, Geometry of Banach Spaces, Proc.Conf.Strobl 1989, London Math. Soc. Lecture Note Series 158, 49–63, Cambridge University Press, (1990)
18. Clarkson, J.A.: Uniformly convex spaces, Trans. Amer. Math. Soc. **40**, 396–414, (1936)
19. Clarkson, J.A., Erdös, P: Approximation by polynomials, Duke Math. J., **10**, 5–11, (1943)
20. Day, M.M.: Normed Linear Spaces, Springer, Berlin Heidelberg New York, (1973)
21. Delbaen, F., Jarchow, H., Pelczynski, A.: Subspaces of L_p isometric to l_p, preprint
22. De Vore, R.A., Lorentz, G.G.: Constructive Approximation, Springer, Berlin Heidelberg New York, (1993)
23. Dor, L.E., Odell, E.: Monotone bases in L_p, Pacific J. Mat. **60**, 51–61, (1975)
24. Dvoretsky, A.: Some results on convex bodies and Banach spaces, Proc. Symp. on Linear Spaces, Jerusalem, (1961)
25. Edelstein, I.S., Wojtaszczyk, P.: On projections and unconditional bases in direct sums of Banach spaces, Studia Math., **56**, 263–276, (1973)
26. Enflo, P.: Banach spaces which can be given an equivalent uniformly convex norm, Israel J. Math., **13**, 281–287, (1972)
27. Enflo, P.: A counterexample to the approximation property in Banach spaces, Acta Math., **130**, 309–317, (1973)
28. Enflo P., Gurariy, V.I.: On Dirichlet sums and Montgomery conjecture, to appear
29. Enflo, P., Gurariy, V.I., Lomonosov, V., Lyubich, Y.: Exponential numbers of linear operators in normed spaces, Linear Algebra and its Applications, **219**, 25–60, (1995)
30. Fonf, V.F.: Polyhedral Banach spaces, Matematicheskie Zametki, **30**, 627–634, (1981)
31. Gelfond, A.O.: Calcul des Differences Finies, Dunod, Paris, (1963)
32. Gohberg, I.Z., Markus, A.S.: Two theorems on the opening of subspaces of Banach spaces, Uspeki Mat. Nauk, **14**,135–140, (1959)
33. von Golitschek, M.: A short proof of Müntz's theorem, J. Approx. Theory, **39**, 394–395, (1983)
34. Goodner, D.B.: Projections in normed linear spaces, Trans. AMS, **69**, 89–108, (1950)
35. Grinblum, M.M.: Some theorems on bases in Banach spaces, Soviet Dokladi, **31**, 428–432, (1941)
36. Grinblum, M.M.: On the representation of a Banach space in a direct sum of subspaces, Soviet Dokladi, **70**, 747–752, (1950)
37. Grünbaum, B.: Projection constants, Trans. AMS, **95**, 451–465, (1960)
38. Gurariy, N.I.: Coefficient sequences in Hilbert and Banach space expansions, Math. USSR Izvestija, **5**, 226–229, (1971)
39. Gurariy, N.I., V.I.Gurariy, V.I.: On a sequential property of the space l, Mat. Issledovaniya Moldav. Acad. Nauk,, 140–143, (1970)
40. Gurariy, N.I., Gurariy, V.I.: On bases in uniformly convex and uniformly smooth Banach spaces, Izvestiya Acad. Nauk SSSR, **35**, 210–215, (1971)
41. Gurariy, V.I.: Bases in spaces of continuous functions, Soviet Dokladi, **148**, 493–495, (1963)

42. Gurariy, V.I.: On some geometrical characteristics of subspaces and bases in Banach spaces, Coll. Math., **13**, 59–63, (1964)

43. Gurariy, V.I.: On opening and inclinations of subspaces in Banach spaces, Theory of Functions, Functional Analysis and Applications, **1**, 194–204, (1965)

44. Gurariy, V.I.: On uniformly convex and uniformly smooth Banach spaces, Theory of Functions, Functional Analysis and Applications, **1**, 205–211, (1965)

45. Gurariy, V.I.: Space of universal disposition, Soviet Dokladi, **163**, 1050–1053, (1965)

46. Gurariy, V.I.: Bases in spaces of continuous functions on compacts and some geometrical problems, Izvestiya Acad. Nauk SSSR, **30**, 289–306, (1966)

47. Gurariy, V.I.: Subspaces and bases in spaces of continuous functions, Soviet Dokladi, **167**, 971–973, (1966)

48. Gurariy, V.I.: Subspaces of differentiable functions, Theory of Functions, Functional Analysis and Applications, 116–119, (1967)

49. Gurariy, V.I.: The projection function of Banach spaces, Doctoral Dissertation, Kharkov, (1973)

50. Gurariy, V.I.: Existence of non-hereditary full systems in separable Banach spaces, Trans. Leningrad Inst. Math., **31** ,(1980)

51. Gurariy, V.I.: Generalization of the Pythagoras-Parseval theorem for all bases in Hilbert space, to appear in Functional Analysis and its Appl.

52. Gurariy, V.I., M.I.Kadec, M.I.: On minimal systems and quasicomplements in Banach spaces, Soviet Dokladi, **145**, 504–506, (1962)

53. Gurariy, V.I., Kadec, M., Macaev, V.: On Banach-Mazur distances between certain Minkowski spaces, Bull. Acad. Polon. Sci., ser. Math., **13** , 573–576, (1965)

54. Gurariy, V.I., Macaev, V.: Lacunary power sequences in C and L_p, Izvestiya Acad. Nauk SSSR, **30**, 3–14, (1966)

55. Gurariy, V.I., Meletidy, M.: The stability of completeness of sequences in Banach spaces, Bull. Acad. Polon. Sci., Math., Mech., Astr., **18**, 533–536, (1970)

56. Gurariy, V.I., Meletidy, M.: On estimating the coefficients of approximating polynomials, Funct. Analysis and its Appl., **5**, Moscow, (1971)

57. Gurariy, V.I., Milman, V.: Direct and opposite problems in sequences theory, International Math. Congress, Moscow, (1966)

58. Hardy, G.H.: Divergent Series, Oxford University Press, (1949)

59. Hardy, G.H., Littlewood, J.E.: A further note on the converse of Abel's theorem, Proc. Lond. Math. Soc., **25**, 219–236, (1926)

60. Hewitt, E., Stromberg, K.: Real and Abstract Analysis, Springer, Berlin Heidelberg New York, (1975)

61. Ivanov, L.A.: Some two-dimensional Banach spaces with symmetric inclination, Funct. Analysis and its Appl., **6**, 66–67, (1968)

62. Johnson, W.B.: Factoring compact operators, Israel J. Math., **9**, 337–345, (1971)

63. Johnson, W.B., Lindenstrauss, J., Editors: Handbook of the Goemetry of Banach Spaces, North-Holland, Elsevier, (2001)

64. Johnson, W.B., Rosenthal, H.P., M.Zippin, M.: On bases, finite dimensional decompositions and weaker structures in Banach spaces, Israel J. Mat **9**, 488–506, (1971)

65. Johnson, W.B., Zippin, W.: On subspaces of quotients of $(\sum G_n)_{l_p}$ and $(\sum G_n)_{c_0}$, Israel J. Math., **13**, 311–316, (1972)

66. Kadec, M.I.: Unconditionally converging series in uniformly convex spaces, Uspeki Mat. Nauk, **11**, 185–190, (1956)
67. Kadec, M.I., Mityagin, B.S.: Complemented subspaces in Banach spaces, Uspeki Mat. Nauk, **28**, 77–94, (1971)
68. Kadec, M.I., Pelczynski, A.: Bases, lacunary sequences and complemented subspaces in L_p, Studia Math., **21**, 161–171, (1962)
69. Kadec, M.I., Snobar, M.G.: Certain functionals on the Minkowski compact, Mat. Zametki, **10**, 453–458, (1971)
70. Kalton, N.J., Werner, D.: Property (M), M-ideals and almost isometric stucture of Banach spaces, Journal Reine Angew. Math., **461**, 137–178 (1995)
71. Kelley, J.L.: Banach spaces with the extension property, Trans. AMS, **72**, 323–326, (1952)
72. Koldobskii, A.L.: Isometries in the spaces $L_p(X; L_q)$ and equimeasurability, Izvestiya Vusov. Mat., **3**, 25–34, (1989)
73. König, H., Lewis, D.L.: A strict inequality for projection constants, J. of Funct. Analysis, **74**, 328–332, (1987)
74. Krasnoselski, M.A., Krein, M.G., Milman, D.P.: On the defect numbers of operators in Banach spaces and some geometric questions, Mat. Inst. of Ukrainian Acad. Sci., **11**, 97–112, (1948)
75. Krein, M.G., Milman, D.P., M.A.Rutman, M.A.: On the stability of bases, Zaiski Khark. Mat. Soc., **14**, 106–110, (1940)
76. Kroó, A., J.Szabados, J.: Müntz-type problems for Bernstein polynomials, J. Approx. Theory, **78**, 446–457, (1994)
77. Lazar, A.J., Lindenstrauss, J.: Banach spaces whose duals are L_1-spaces and their representing matrices, Acta Math., **126**, 165–194, (1971)
78. Linde, W.: Uniqueness theorems for measures in L_r and $C_0(\Omega)$, Math. Ann., **174**, 617–626, (1986)
79. Lindenstrauss, J.: On the modulus of smoothness and divergent series in Banach spaces, Mich. Math. J., **10**, 241–252, (1963)
80. Lindenstrauss, J.: Extension of compact operators, Mem. Amer. Math. Soc. No 48, (1964)
81. Lindenstrauss, J., Milman, V.: Local Theory of Normed Spaces and its Applications to Convexity, Elsevier Science Publisher, (1993)
82. Lindenstrauss, J., Rosenthal, H.P.: The $\mathcal{L}_p$–spaces, Israel J. Math., **7**, 325–349, (1969)
83. Lindenstrauss, J., Tzafriri, L.: Classical Banach Spaces, Lecture Notes 338, Springer, Berlin Heidelberg New York,(1973)
84. Lindenstrauss, J., Tzafriri, L.: Classical Banach Spaces I and II, Springer, Berlin Heidelberg New York, (1977)
85. Lorentz, G.G.: Approximation by incomplete polynomials (problems and results), Padé and Rational Approximations: Theory and Applications, (E.B.Saff and R.S.Varga, eds.), 289–302, Academic Press Inc., New York, (1977)
86. Lusky, W.: The Gurarij-spaces are unique, Arch.Math. **27**, 627–635, (1976)
87. Lusky, W.: Separable Lindenstrauss-spaces, in Functional Analysis: Surveys and Recent Results, 15–28, North Holland, (1977)
88. Lusky, W.: On separable Lindenstrauss-spaces, J. of Funct. Analysis **26**, 103–120, (1977)
89. Lusky, W.: Some Consequences of Rudin's Paper "L_p-Isometries and Equimeasurability, Indiana University Math. J., **27**, 859–866, (1978)

90. Lusky, W.: A note on the Paper "The Poulsen Simplex" of Lindenstrauss, Olsen and Sternfeld, Ann. l'Inst. Fourier (Grenoble), **28**, 233–243, (1978)

91. Lusky, W.: A note on rotations in separable Banach spaces, Studia Math., **65**, 239–242, (1979)

92. Lusky, W.: On a construction of Lindenstrauss and Wulbert, J. of Funct. Analysis **31**, 42–51, (1979)

93. Lusky, W.: On the primariness of the Poulsen simplex space, Israel J. Math., **37**, 151- 163, (1980)

94. Lusky, W.: A note on Banach spaces containing c_0 or C_∞, J. of Funct. Analysis, **62**, 1–7, (1985)

95. Lusky, W.: On a class of interpolating bases, Seminario Matematico Garcia de Galdeano, Universidad de Zaragoza, 72–82, (1985)

96. Lusky, W.: On non-separable simplex spaces, Math. Scand. **61**, 276–285, (1987)

97. Lusky, W.: A note on interpolating bases, J. of Approx. Theory, **52**, 107–118, (1988)

98. Lusky, W.: Every L_1- predual is complemented in a simplex space, Israel J. Math., **64**, 169–178, (1988)

99. Lusky, W.: On Banach spaces with the commuting bounded approximation property, Arch. Math., **58**, 568–574 (1992)

100. Lusky, W.: On Banach spaces with bases, J. of Funct. Analysis, **138**, 410–425 (1996)

101. Lusky, W.: Three space properties and basis extensions, Israel J. Math., **107**, 17–27, (1998)

102. Markushevich, A.I.: On bases (in wide sense), Soviet Dokladi, **41**, 241–243, (1943)

103. Mazur, S.: Über konvexe Mengen in linearen normierten Räumen, Studia Math., **4**, 70–84, (1933)

104. Mazur, S.: On the convergence of lacunary polynomials, Studia Math., **89**, 75–78, (1988)

105. McGehee, O.C., Pigno, L., Smith, B.: Hardy's inequality and the L^1-norm of exponential sums, Ann. Math., **113**, 613–618, (1981)

106. Milman, D.P.: Some criteria of regularity of Banach spaces, Soviet Dokladi, 243–246, (1938)

107. Milman, V.: Geometric theory of Banach spaces II, Uspeki Mat. Nauk., **26**, 73–149, (1971)

108. Milutin, A.A.: Isomorphisms of spaces of continuous functions on compacts of continuum power, Theory of Functions, Functional Analysis and Applications, **2**, 150–156, (1966)

109. Muckenhoupt, B.: Mean convergence of Jacobi series, Proc. Amer. Math. Soc., **23**, 306–310, (1969)

110. Müntz, C.: Über den Approximationssatz von Weierstraß , H.A.Schwarz Festschrift, Berlin, 303–312, (1914)

111. Nachbin, L.: A theorem of Hahn-Banach type for linear transformations, Trans. AMS, **68**, 28–46, (1950)

112. Natanson, I.P.: Constructive Function Theory, Vol. 1, Frederick Ungar, New York, (1964)

113. Olevskii, A.M.: Fourier series of continuous functions with respect to bounded orthonormal systems, Izv. Akad. Nauk. SSSR Ser. Mat., **30**, 387–432, (1966)

114. Ovsepian, R.I., Pelczynski, A.: The existence in every separable Banach space of a fundamental total and bounded biorthogonal sequence and related constructions of uniformly bounded orthonormal systems in L^2, Studia Math., **54**, 149–159, (1975)

115. Paley, R.E., Wiener, N.: Fourier transforms in the complex domain, Amer. Math. Soc. Coll. Publ., **19**, (1934)

116. Pelczynski, A.: Projections in certain Banach spaces, Studia Math., **19**, 209–228, (1960)

117. Pelczynski, A.: Universal bases, Studia Math., **32** , 247–268, (1969)

118. Pelczynski, A.: Any separable Banach space with the bounded approximation property is a complemented subspace of a Banach space with bases, Studia Math., **40**, 239–242, (1971)

119. Pelczynski, A., Rosenthal, H.P.: Localization techniques in L^p-spaces, Studia Math., **52**, 263–289, (1975)

120. Pietsch, A.: Operator Ideals, Verlag der Wissenschaften, Berlin, (1978)

121. Pisier, G.: Counterexamples to a conjecture of Grothendieck, Acta Math., **151**, 181–208, (1983)

122. Plotkin, A.I.: Continuation of L_p-isometries, J. Sov. Math., **2**, 143–165, (1974)

123. Read, C.J.: Different forms of the approximation property, to appear

124. Rosenthal, H.P.: On totally incomparable Banach spaces, J. Funct. Analysis, **4**, 167–175, (1969)

125. Rudin, W.: L_p-isometries and equimeasurability, Indiana Univ. Math. J., **25**, 215–228, (1976)

126. Rutovitz, D.: Some parameters associated with finite dimensional Banach spaces, J. London Math. Soc., **40**, 241–255, (1965)

127. Saff, E.B., Varga, R.S.: On lacunary incomplete polynomials, Math. Z., **177**, 297–314, (1981)

128. Schwartz, L.: Étude des sommes d'exponentielles, Hermann, Paris, (1959)

129. Singer, I.: Bases in Banach Spaces I, Springer, Berlin Heidelberg New York, (1970)

130. Shekhtman, B.: On the norms of some projections, Lecture Notes in Math. 166, Springer, Berlin Heidelberg New York, 177–185, (1984)

131. Shekhtman, B.: On the norms of interpolating operators, Israel J. Math., **64**, 39–48, (1988)

132. Shekhtman, B.: Some examples concerning projection constants, Approximation theory, spline functions and its applications, (Maratea, 1991), Nato Adv. Sci. Inst. Ser. C Math. Phys. Sci. 356, Kluwer, Dordrecht, 471–476, (1992)

133. Šmul'yan, V.L.: Sur la structure de la sphère unitaire dans l'espace de Banach, Math. USSR Sbornik, **51**, 545–561, (1941)

134. Szarek, S.J.: Bases and biorthogonal systems in the spaces C and L_1, Arkiv for Matematik, **17**, 255–271, (1979)

135. Szarek, S.J.: A Banach space without a basis which has the bounded approximation property, Acta Math., **159**, 81–98, (1987)

136. Szasz, O.: Über die Approximation stetiger Funktionen durch lineare Aggregate von Potenzen, Math. Ann., **77**, 482–496, (1916)

137. Terenzi, P.: Every norming M-basis of a separable Banach space has a block perturbation which is a norming strong M-basis, Extracta Mathematica, 161–169, (1990)

138. Tomczak-Jaegermann, N.: Banach-Mazur Distances and Finite-Dimensional Operator Ideals, Pitman Monographs, New York, (1989)

139. Wojcieszyk, B.: Some geometric and analytic properties of Müntz spaces, Preprint, (1994)

140. Wojciechowski, M.: On the convergence of lacunary polynomials, preprint, (1995)

141. Wojtaszczyk, P.: Some remarks on the Gurarij space, Studia Math., **41**, 207–210, (1972)

142. Wojtaszczyk, P.: The Franklin system is an unconditional basis in H^1, Arkiv för Math., **20**, 293–300, (1982)

143. Wojtaszczyk, P.: Banach spaces for analysts, Cambridge University Press, (1991)

144. Wojtaszczyk, P., Wozniakowski, K.: Orthonormal polynomial bases in function spaces, Inst. of Math., Polish Ac. of Sc., Preprint 475, (1990)

145. Zippin, M.: On finite-dimensional $\mathcal{P}_\lambda$-spaces with small λ, Israel J. Math., **39**, 359–365, (1981)

Index

Lecture Notes in Mathematics

For information about earlier volumes
please contact your bookseller or Springer
LNM Online archive: springerlink.com

Vol. 1722: R. McCutcheon, Elemental Methods in Ergodic Ramsey Theory (1999)

Vol. 1723: J. P. Croisille, C. Lebeau, Diffraction by an Immersed Elastic Wedge (1999)

Vol. 1724: V. N. Kolokoltsov, Semiclassical Analysis for Diffusions and Stochastic Processes (2000)

Vol. 1725: D. A. Wolf-Gladrow, Lattice-Gas Cellular Automata and Lattice Boltzmann Models (2000)

Vol. 1726: V. Marić, Regular Variation and Differential Equations (2000)

Vol. 1727: P. Kravanja M. Van Barel, Computing the Zeros of Analytic Functions (2000)

Vol. 1728: K. Gatermann Computer Algebra Methods for Equivariant Dynamical Systems (2000)

Vol. 1729: J. Azéma, M. Émery, M. Ledoux, M. Yor (Eds.) Séminaire de Probabilités XXXIV (2000)

Vol. 1730: S. Graf, H. Luschgy, Foundations of Quantization for Probability Distributions (2000)

Vol. 1731: T. Hsu, Quilts: Central Extensions, Braid Actions, and Finite Groups (2000)

Vol. 1732: K. Keller, Invariant Factors, Julia Equivalences and the (Abstract) Mandelbrot Set (2000)

Vol. 1733: K. Ritter, Average-Case Analysis of Numerical Problems (2000)

Vol. 1734: M. Espedal, A. Fasano, A. Mikelić, Filtration in Porous Media and Industrial Applications. Cetraro 1998. Editor: A. Fasano. 2000.

Vol. 1735: D. Yafaev, Scattering Theory: Some Old and New Problems (2000)

Vol. 1736: B. O. Turesson, Nonlinear Potential Theory and Weighted Sobolev Spaces (2000)

Vol. 1737: S. Wakabayashi, Classical Microlocal Analysis in the Space of Hyperfunctions (2000)

Vol. 1738: M. Émery, A. Nemirovski, D. Voiculescu, Lectures on Probability Theory and Statistics (2000)

Vol. 1739: R. Burkard, P. Deuflhard, A. Jameson, J.-L. Lions, G. Strang, Computational Mathematics Driven by Industrial Problems. Martina Franca, 1999. Editors: V. Capasso, H. Engl, J. Periaux (2000)

Vol. 1740: B. Kawohl, O. Pironneau, L. Tartar, J.-P. Zolesio, Optimal Shape Design. Tróia, Portugal 1999. Editors: A. Cellina, A. Ornelas (2000)

Vol. 1741: E. Lombardi, Oscillatory Integrals and Phenomena Beyond all Algebraic Orders (2000)

Vol. 1742: A. Unterberger, Quantization and Non-holomorphic Modular Forms (2000)

Vol. 1743: L. Habermann, Riemannian Metrics of Constant Mass and Moduli Spaces of Conformal Structures (2000)

Vol. 1744: M. Kunze, Non-Smooth Dynamical Systems (2000)

Vol. 1745: V. D. Milman, G. Schechtman (Eds.), Geometric Aspects of Functional Analysis. Israel Seminar 1999-2000 (2000)

Vol. 1746: A. Degtyarev, I. Itenberg, V. Kharlamov, Real Enriques Surfaces (2000)

Vol. 1747: L. W. Christensen, Gorenstein Dimensions (2000)

Vol. 1748: M. Ruzicka, Electrorheological Fluids: Modeling and Mathematical Theory (2001)

Vol. 1749: M. Fuchs, G. Seregin, Variational Methods for Problems from Plasticity Theory and for Generalized Newtonian Fluids (2001)

Vol. 1750: B. Conrad, Grothendieck Duality and Base Change (2001)

Vol. 1751: N. J. Cutland, Loeb Measures in Practice: Recent Advances (2001)

Vol. 1752: Y. V. Nesterenko, P. Philippon, Introduction to Algebraic Independence Theory (2001)

Vol. 1753: A. I. Bobenko, U. Eitner, Painlevé Equations in the Differential Geometry of Surfaces (2001)

Vol. 1754: W. Bertram, The Geometry of Jordan and Lie Structures (2001)

Vol. 1755: J. Azéma, M. Émery, M. Ledoux, M. Yor (Eds.), Séminaire de Probabilités XXXV (2001)

Vol. 1756: P. E. Zhidkov, Korteweg de Vries and Nonlinear Schrödinger Equations: Qualitative Theory (2001)

Vol. 1757: R. R. Phelps, Lectures on Choquet's Theorem (2001)

Vol. 1758: N. Monod, Continuous Bounded Cohomology of Locally Compact Groups (2001)

Vol. 1759: Y. Abe, K. Kopfermann, Toroidal Groups (2001)

Vol. 1760: D. Filipović, Consistency Problems for Heath-Jarrow-Morton Interest Rate Models (2001)

Vol. 1761: C. Adelmann, The Decomposition of Primes in Torsion Point Fields (2001)

Vol. 1762: S. Cerrai, Second Order PDE's in Finite and Infinite Dimension (2001)

Vol. 1763: J.-L. Loday, A. Frabetti, F. Chapoton, F. Goichot, Dialgebras and Related Operads (2001)

Vol. 1764: A. Cannas da Silva, Lectures on Symplectic Geometry (2001)

Vol. 1765: T. Kerler, V. V. Lyubashenko, Non-Semisimple Topological Quantum Field Theories for 3-Manifolds with Corners (2001)

Vol. 1766: H. Hennion, L. Hervé, Limit Theorems for Markov Chains and Stochastic Properties of Dynamical Systems by Quasi-Compactness (2001)

Vol. 1767: J. Xiao, Holomorphic Q Classes (2001)

Vol. 1768: M.J. Pflaum, Analytic and Geometric Study of Stratified Spaces (2001)

Vol. 1769: M. Alberich-Carramiñana, Geometry of the Plane Cremona Maps (2002)

Vol. 1770: H. Gluesing-Luerssen, Linear Delay-Differential Systems with Commensurate Delays: An Algebraic Approach (2002)

Vol. 1771: M. Émery, M. Yor (Eds.), Séminaire de Probabilités 1967-1980. A Selection in Martingale Theory (2002)

Vol. 1772: F. Burstall, D. Ferus, K. Leschke, F. Pedit, U. Pinkall, Conformal Geometry of Surfaces in S^4 (2002)

Vol. 1773: Z. Arad, M. Muzychuk, Standard Integral Table Algebras Generated by a Non-real Element of Small Degree (2002)

Vol. 1774: V. Runde, Lectures on Amenability (2002)

Vol. 1775: W. H. Meeks, A. Ros, H. Rosenberg, The Global Theory of Minimal Surfaces in Flat Spaces. Martina Franca 1999. Editor: G. P. Pirola (2002)

Vol. 1776: K. Behrend, C. Gomez, V. Tarasov, G. Tian, Quantum Comohology. Cetraro 1997. Editors: P. de Bartolomeis, B. Dubrovin, C. Reina (2002)

Vol. 1777: E. García-Río, D. N. Kupeli, R. Vázquez-Lorenzo, Osserman Manifolds in Semi-Riemannian Geometry (2002)

Vol. 1778: H. Kiechle, Theory of K-Loops (2002)

Vol. 1779: I. Chueshov, Monotone Random Systems (2002)

Vol. 1780: J. H. Bruinier, Borcherds Products on O(2,1) and Chern Classes of Heegner Divisors (2002)

Vol. 1781: E. Bolthausen, E. Perkins, A. van der Vaart, Lectures on Probability Theory and Statistics. Ecole d' Eté de Probabilités de Saint-Flour XXIX-1999. Editor: P. Bernard (2002)

Vol. 1782: C.-H. Chu, A. T.-M. Lau, Harmonic Functions on Groups and Fourier Algebras (2002)

Vol. 1783: L. Grüne, Asymptotic Behavior of Dynamical and Control Systems under Perturbation and Discretization (2002)

Vol. 1784: L.H. Eliasson, S. B. Kuksin, S. Marmi, J.-C. Yoccoz, Dynamical Systems and Small Divisors. Cetraro, Italy 1998. Editors: S. Marmi, J.-C. Yoccoz (2002)

Vol. 1785: J. Arias de Reyna, Pointwise Convergence of Fourier Series (2002)

Vol. 1786: S. D. Cutkosky, Monomialization of Morphisms from 3-Folds to Surfaces (2002)

Vol. 1787: S. Caenepeel, G. Militaru, S. Zhu, Frobenius and Separable Functors for Generalized Module Categories and Nonlinear Equations (2002)

Vol. 1788: A. Vasil'ev, Moduli of Families of Curves for Conformal and Quasiconformal Mappings (2002)

Vol. 1789: Y. Sommerhäuser, Yetter-Drinfel'd Hopf algebras over groups of prime order (2002)

Vol. 1790: X. Zhan, Matrix Inequalities (2002)

Vol. 1791: M. Knebusch, D. Zhang, Manis Valuations and Prüfer Extensions I: A new Chapter in Commutative Algebra (2002)

Vol. 1792: D. D. Ang, R. Gorenflo, V. K. Le, D. D. Trong, Moment Theory and Some Inverse Problems in Potential Theory and Heat Conduction (2002)

Vol. 1793: J. Cortés Monforte, Geometric, Control and Numerical Aspects of Nonholonomic Systems (2002)

Vol. 1794: N. Pytheas Fogg, Substitution in Dynamics, Arithmetics and Combinatorics. Editors: V. Berthé, S. Ferenczi, C. Mauduit, A. Siegel (2002)

Vol. 1795: H. Li, Filtered-Graded Transfer in Using Noncommutative Gröbner Bases (2002)

Vol. 1796: J.M. Melenk, hp-Finite Element Methods for Singular Perturbations (2002)

Vol. 1797: B. Schmidt, Characters and Cyclotomic Fields in Finite Geometry (2002)

Vol. 1798: W.M. Oliva, Geometric Mechanics (2002)

Vol. 1799: H. Pajot, Analytic Capacity, Rectifiability, Menger Curvature and the Cauchy Integral (2002)

Vol. 1800: O. Gabber, L. Ramero, Almost Ring Theory (2003)

Vol. 1801: J. Azéma, M. Émery, M. Ledoux, M. Yor (Eds.), Séminaire de Probabilités XXXVI (2003)

Vol. 1802: V. Capasso, E. Merzbach, B.G. Ivanoff, M. Dozzi, R. Dalang, T. Mountford, Topics in Spatial Stochastic Processes. Martina Franca, Italy 2001. Editor: E. Merzbach (2003)

Vol. 1803: G. Dolzmann, Variational Methods for Crystalline Microstructure – Analysis and Computation (2003)

Vol. 1804: I. Cherednik, Ya. Markov, R. Howe, G. Lusztig, Iwahori-Hecke Algebras and their Representation Theory. Martina Franca, Italy 1999. Editors: V. Baldoni, D. Barbasch (2003)

Vol. 1805: F. Cao, Geometric Curve Evolution and Image Processing (2003)

Vol. 1806: H. Broer, I. Hoveijn. G. Lunther, G. Vegter, Bifurcations in Hamiltonian Systems. Computing Singularities by Gröbner Bases (2003)

Vol. 1807: V. D. Milman, G. Schechtman (Eds.), Geometric Aspects of Functional Analysis. Israel Seminar 2000-2002 (2003)

Vol. 1808: W. Schindler, Measures with Symmetry Properties (2003)

Vol. 1809: O. Steinbach, Stability Estimates for Hybrid Coupled Domain Decomposition Methods (2003)

Vol. 1810: J. Wengenroth, Derived Functors in Functional Analysis (2003)

Vol. 1811: J. Stevens, Deformations of Singularities (2003)

Vol. 1812: L. Ambrosio, K. Deckelnick, G. Dziuk, M. Mimura, V. A. Solonnikov, H. M. Soner, Mathematical Aspects of Evolving Interfaces. Madeira, Funchal, Portugal 2000. Editors: P. Colli, J. F. Rodrigues (2003)

Vol. 1813: L. Ambrosio, L. A. Caffarelli, Y. Brenier, G. Buttazzo, C. Villani, Optimal Transportation and its Applications. Martina Franca, Italy 2001. Editors: L. A. Caffarelli, S. Salsa (2003)

Vol. 1814: P. Bank, F. Baudoin, H. Föllmer, L.C.G. Rogers, M. Soner, N. Touzi, Paris-Princeton Lectures on Mathematical Finance 2002 (2003)

Vol. 1815: A. M. Vershik (Ed.), Asymptotic Combinatorics with Applications to Mathematical Physics. St. Petersburg, Russia 2001 (2003)

Vol. 1816: S. Albeverio, W. Schachermayer, M. Talagrand, Lectures on Probability Theory and Statistics. Ecole d'Eté de Probabilités de Saint-Flour XXX-2000. Editor: P. Bernard (2003)

Vol. 1817: E. Koelink, W. Van Assche(Eds.), Orthogonal Polynomials and Special Functions. Leuven 2002 (2003)

Vol. 1818: M. Bildhauer, Convex Variational Problems with Linear, nearly Linear and/or Anisotropic Growth Conditions (2003)

Vol. 1819: D. Masser, Yu. V. Nesterenko, H. P. Schlickewei, W. M. Schmidt, M. Waldschmidt, Diophantine Approximation. Cetraro, Italy 2000. Editors: F. Amoroso, U. Zannier (2003)

Vol. 1820: F. Hiai, H. Kosaki, Means of Hilbert Space Operators (2003)

Vol. 1821: S. Teufel, Adiabatic Perturbation Theory in Quantum Dynamics (2003)

Vol. 1822: S.-N. Chow, R. Conti, R. Johnson, J. Mallet-Paret, R. Nussbaum, Dynamical Systems. Cetraro, Italy 2000. Editors: J. W. Macki, P. Zecca (2003)

Vol. 1823: A. M. Anile, W. Allegretto, C. Ringhofer, Mathematical Problems in Semiconductor Physics. Cetraro, Italy 1998. Editor: A. M. Anile (2003)

Vol. 1824: J. A. Navarro González, J. B. Sancho de Salas, $\mathscr{C}^{\infty}$ – Differentiable Spaces (2003)

Vol. 1825: J. H. Bramble, A. Cohen, W. Dahmen, Multiscale Problems and Methods in Numerical Simulations, Martina Franca, Italy 2001. Editor: C. Canuto (2003)

Vol. 1826: K. Dohmen, Improved Bonferroni Inequalities via Abstract Tubes. Inequalities and Identities of Inclusion-Exclusion Type. VIII, 113 p, 2003.

Vol. 1827: K. M. Pilgrim, Combinations of Complex Dynamical Systems. IX, 118 p, 2003.

Vol. 1828: D. J. Green, Gröbner Bases and the Computation of Group Cohomology. XII, 138 p, 2003.

Vol. 1829: E. Altman, B. Gaujal, A. Hordijk, Discrete-Event Control of Stochastic Networks: Multimodularity and Regularity. XIV, 313 p, 2003.

Vol. 1830: M. I. Gil', Operator Functions and Localization of Spectra. XIV, 256 p, 2003.

Vol. 1831: A. Connes, J. Cuntz, E. Guentner, N. Higson, J. E. Kaminker, Noncommutative Geometry, Martina Franca, Italy 2002. Editors: S. Doplicher, L. Longo (2004)

Vol. 1832: J. Azéma, M. Émery, M. Ledoux, M. Yor (Eds.), Séminaire de Probabilités XXXVII (2003)

Vol. 1833: D.-Q. Jiang, M. Qian, M.-P. Qian, Mathematical Theory of Nonequilibrium Steady States. On the Frontier of Probability and Dynamical Systems. IX, 280 p, 2004.

Vol. 1834: Yo. Yomdin, G. Comte, Tame Geometry with Application in Smooth Analysis. VIII, 186 p, 2004.

Vol. 1835: O.T. Izhboldin, B. Kahn, N.A. Karpenko, A. Vishik, Geometric Methods in the Algebraic Theory of Quadratic Forms. Summer School, Lens, 2000. Editor: J.-P. Tignol (2004)

Vol. 1836: C. Năstăsescu, F. Van Oystaeyen, Methods of Graded Rings. XIII, 304 p, 2004.

Vol. 1837: S. Tavaré, O. Zeitouni, Lectures on Probability Theory and Statistics. Ecole d'Eté de Probabilités de Saint-Flour XXXI-2001. Editor: J. Picard (2004)

Vol. 1838: A.J. Ganesh, N.W. O'Connell, D.J. Wischik, Big Queues. XII, 254 p, 2004.

Vol. 1839: R. Gohm, Noncommutative Stationary Processes. VIII, 170 p, 2004.

Vol. 1840: B. Tsirelson, W. Werner, Lectures on Probability Theory and Statistics. Ecole d'Eté de Probabilités de Saint-Flour XXXII-2002. Editor: J. Picard (2004)

Vol. 1841: W. Reichel, Uniqueness Theorems for Variational Problems by the Method of Transformation Groups (2004)

Vol. 1842: T. Johnsen, A.L. Knutsen, K3 Projective Models in Scrolls (2004)

Vol. 1843: B. Jefferies, Spectral Properties of Noncommuting Operators (2004)

Vol. 1844: K.F. Siburg, The Principle of Least Action in Geometry and Dynamics (2004)

Vol. 1845: Min Ho Lee, Mixed Automorphic Forms, Torus Bundles, and Jacobi Forms (2004)

Vol. 1846: H. Ammari, H. Kang, Reconstruction of Small Inhomogeneities from Boundary Measurements (2004)

Vol. 1847: T.R. Bielecki, T. Björk, M. Jeanblanc, M. Rutkowski, J.A. Scheinkman, W. Xiong, Paris-Princeton Lectures on Mathematical Finance 2003 (2004)

Vol. 1848: M. Abate, J. E. Fornaess, X. Huang, J. P. Rosay, A. Tumanov, Real Methods in Complex and CR Geometry, Martina Franca, Italy 2002. Editors: D. Zaitsev, G. Zampieri (2004)

Vol. 1849: Martin L. Brown, Heegner Modules and Elliptic Curves (2004)

Vol. 1850: V. D. Milman, G. Schechtman (Eds.), Geometric Aspects of Functional Analysis. Israel Seminar 2002-2003 (2004)

Vol. 1851: O. Catoni, Statistical Learning Theory and Stochastic Optimization (2004)

Vol. 1852: A.S. Kechris, B.D. Miller, Topics in Orbit Equivalence (2004)

Vol. 1853: Ch. Favre, M. Jonsson, The Valuative Tree (2004)

Vol. 1854: O. Saeki, Topology of Singular Fibers of Differential Maps (2004)

Vol. 1855: G. Da Prato, P.C. Kunstmann, I. Lasiecka, A. Lunardi, R. Schnaubelt, L. Weis, Functional Analytic Methods for Evolution Equations. Editors: M. Iannelli, R. Nagel, S. Piazzera (2004)

Vol. 1856: K. Back, T.R. Bielecki, C. Hipp, S. Peng, W. Schachermayer, Stochastic Methods in Finance, Bressanone/Brixen, Italy, 2003. Editors: M. Fritelli, W. Runggaldier (2004)

Vol. 1857: M. Émery, M. Ledoux, M. Yor (Eds.), Séminaire de Probabilités XXXVIII (2005)

Vol. 1858: A.S. Cherny, H.-J. Engelbert, Singular Stochastic Differential Equations (2005)

Vol. 1859: E. Letellier, Fourier Transforms of Invariant Functions on Finite Reductive Lie Algebras (2005)

Vol. 1860: A. Borisyuk, G.B. Ermentrout, A. Friedman, D. Terman, Tutorials in Mathematical Biosciences I. Mathematical Neurosciences (2005)

Vol. 1861: G. Benettin, J. Henrard, S. Kuksin, Hamiltonian Dynamics – Theory and Applications, Cetraro, Italy, 1999. Editor: A. Giorgilli (2005)

Vol. 1862: B. Helffer, F. Nier, Hypoelliptic Estimates and Spectral Theory for Fokker-Planck Operators and Witten Laplacians (2005)

Vol. 1863: H. Fürh, Abstract Harmonic Analysis of Continuous Wavelet Transforms (2005)

Vol. 1864: K. Efstathiou, Metamorphoses of Hamiltonian Systems with Symmetries (2005)

Vol. 1865: D. Applebaum, B.V. R. Bhat, J. Kustermans, J. M. Lindsay, Quantum Independent Increment Processes I. From Classical Probability to Quantum Stochastic Calculus. Editors: M. Schürmann, U. Franz (2005)

Vol. 1866: O.E. Barndorff-Nielsen, U. Franz, R. Gohm, B. Kümmerer, S. Thorbjønsen, Quantum Independent Increment Processes II. Structure of Quantum Levy Processes, Classical Probability, and Physics. Editors: M. Schürmann, U. Franz, (2005)

Vol. 1867: J. Sneyd (Ed.), Tutorials in Mathematical Biosciences II. Mathematical Modeling of Calcium Dynamics and Signal Transduction. (2005)

Vol. 1868: A. Jorgenson, S. Lang, Pos_n(R) and Eisenstein Sereies. (2005)

Vol. 1869: A. Dembo, T. Funaki, Lectures on Probability Theory and Statistics. Ecole d'Eté de Probabilités de Saint-Flour XXXIII-2003. Editor: J. Picard (2005)

Vol. 1870: V.I. Gurariy, W. Lusky, Geometry of Müntz Spaces and Related Questions. (2005)

Recent Reprints and New Editions

Vol. 1200: V. D. Milman, G. Schechtman (Eds.), Asymptotic Theory of Finite Dimensional Normed Spaces. 1986. – Corrected Second Printing (2001)

Vol. 1471: M. Courtieu, A.A. Panchishkin, Non-Archimedean L-Functions and Arithmetical Siegel Modular Forms. – Second Edition (2003)

Vol. 1618: G. Pisier, Similarity Problems and Completely Bounded Maps. 1995 – Second, Expanded Edition (2001)

Vol. 1629: J.D. Moore, Lectures on Seiberg-Witten Invariants. 1997 – Second Edition (2001)

Vol. 1638: P. Vanhaecke, Integrable Systems in the realm of Algebraic Geometry. 1996 – Second Edition (2001)

Vol. 1702: J. Ma, J. Yong, Forward-Backward Stochastic Differential Equations and their Applications. 1999. – Corrected 3rd printing (2005)